COMPUTATIONAL FLUID DYNAMICS APPLIED TO WASTE-TO-ENERGY PROCESSES

T0311933

COMPUTATIONAL FLUID DYNAMICS APPLIED TO WASTE-TO-ENERGY PROCESSES

A Hands-On Approach

VALTER BRUNO REIS E. SILVA

Renewable Energy, Polytechnic Institute of Portalegre, Portalegre, Portugal

JOÃO CARDOSO

Instituto Superior Técnico, Universidade de Lisboa, Lisboa, Portugal
Polytechnic Institute of Portalegre, Portalegre, Portugal

Butterworth-Heinemann
An imprint of Elsevier

Butterworth-Heinemann is an imprint of Elsevier
The Boulevard, Langford Lane, Kidlington, Oxford OX5 1GB, United Kingdom
50 Hampshire Street, 5th Floor, Cambridge, MA 02139, United States

© 2020 Elsevier Inc. All rights reserved.

No part of this publication may be reproduced or transmitted in any form or by any means, electronic or mechanical, including photocopying, recording, or any information storage and retrieval system, without permission in writing from the publisher. Details on how to seek permission, further information about the Publisher's permissions policies and our arrangements with organizations such as the Copyright Clearance Center and the Copyright Licensing Agency, can be found at our website: www.elsevier.com/permissions.

This book and the individual contributions contained in it are protected under copyright by the Publisher (other than as may be noted herein).

Notices
Knowledge and best practice in this field are constantly changing. As new research and experience broaden our understanding, changes in research methods, professional practices, or medical treatment may become necessary.

Practitioners and researchers must always rely on their own experience and knowledge in evaluating and using any information, methods, compounds, or experiments described herein. In using such information or methods they should be mindful of their own safety and the safety of others, including parties for whom they have a professional responsibility.

To the fullest extent of the law, neither the Publisher nor the authors, contributors, or editors, assume any liability for any injury and/or damage to persons or property as a matter of products liability, negligence or otherwise, or from any use or operation of any methods, products, instructions, or ideas contained in the material herein.

Library of Congress Cataloging-in-Publication Data
A catalog record for this book is available from the Library of Congress

British Library Cataloguing-in-Publication Data
A catalogue record for this book is available from the British Library

ISBN: 978-0-12-817540-8

For information on all Butterworth-Heinemann publications
visit our website at https://www.elsevier.com/books-and-journals

Publisher: Joe Hayton
Acquisitions Editor: Peter Adamson
Editorial Project Manager: Chris Hockaday
Production Project Manager: Sojan P. Pazhayattil
Cover Designer: Greg Harris

Typeset by SPi Global, India

To my wife, who made me to accept this challenge and is the cornerstone of my life.

Dr. Valter Bruno Reis E. Silva

Contents

Section III   Gasification modeling

Preface

Computational fluid dynamics is not a simple subject! The equations governing the behavior of fluids are the result of over a century of hard and intense work. To make matters worse, complex mathematical tools are required to handle with fluid behavior peculiarities. Consequently, students and even professionals are often overwhelmed in their attempt to connect their mathematical knowledge to practical applications. It is not surprising to find many students looking for help in different internet forums and complaining about sparse or missing information. They are looking for the better strategy to implement in each particular case or asking how they build customized codes. When fluid dynamics courses are taught in universities, their study plans are essentially focused on theoretical approaches meaning that the adequate proficiency and confidence to deal with computational approaches is sometimes neglected.

The major question is how students and professionals become good users in such a complex software. Generally, they face only two valuable options: self-learning based on a trial-error method that is very demanding and time consuming or enrolling in a specialized course that could be very expensive and lengthy. Under such circumstances, a book with simplified theoretical background and worked examples with the necessary analysis of each step comprising problem formulation is a tool required for many. The book combines an adequate level of mathematical background with real-world examples including combustors and gasifiers. By working through examples and taking computer screenshots, applying step-by-step guidelines, and with some customization capabilities, readers will learn to move beyond button pushing and start thinking as professionals.

The use of worked examples in complex processes such as combustion or gasification involving fluid dynamics, heat and mass transfer, and complex chemistry sets allow the readers comprehensive knowledge that can be used in so many other real problems. More specifically, this is also a relevant contribution for students and professionals engaged with thermochemical conversion processes in research and industrial environments. In a broad spectrum, this book can be used by all the individuals interested in multiphase problems, fluid mechanics, simulation, and hydrodynamics.

This book is organized into several chapters balancing theory and application including step-by-step tutorials and all the skills needed to perform real-life simulation calculations.

Acknowledgments

I am indebted to my entire research team for their efforts in all stages of the book and valuable suggestions during the whole process. Special thanks to Mr. João Cardoso, whom I invited to be coauthor of this book, for his relentless effort in supporting me by designing most pictures of this book and playing a very important role in the tutorial chapters, making them more accessible to all users.

I also owe a debt of gratitude to Ms. Raquel Zanol and Ms. Joanne Collett for their outstanding support and help in bringing out the book in the present form.

Finally, my special thanks to the Elsevier team for their concern and valuable support to make this book possible.

CFD workflow implementation

CHAPTER 1

Introduction and overview of using computational fluid dynamics tools

1. Introduction

Waste-to-energy (WtE) is the process of converting waste into electricity and heat, or the path of turning waste into a fuel source [1]. Rapidly increasing urbanization and world population growth mean significantly larger amounts of waste and a greater demand for different sources of energy. WtE strategies are being set to tackle both problems simultaneously, by helping to dispose of the waste while generating an important contribution on the steep energy demand. Waste holds a large potential as a source of renewable energy and greenhouse gases emission reduction but its use is advancing at a slow pace, and beyond several concerted strategies needed by private and public institutions to convince the public opinion and overcome political barriers, jumps in technical features are still required [2].

WtE companies are struggling with major challenges: intensified generalized competition, the event of disruptive technologies, strict government regulations, and some immature technological features. Such issues outline the complexity and the risks for even the best well-prepared players. Beyond the impact of using effective business models, well succeeding companies must embrace new approaches to improve the process design, minimize technological uncertainties, and optimize the process outputs with a reduced number of failed attempts. The ultimate goal of a WtE solution preconizes a reduced environmental footprint combined with a large energy efficiency process and minimum by-products [3].

Reliable quantitative analysis requires a very expensive large-scale experimentation with work developed on laboratory level generating results often far from the reality [4]. Over the last decades, with the increasing computational power and numerical solvers efficiency, computational fluid dynamics (CFD) is broadly used to design, optimize, and predict the physical-chemical phenomena regarding several processes and more recently has been introduced to WtE systems [5]. CFD comprises a set of elaborate mathematical models governed by partial differential equations representing conservation laws for mass, momentum, and energy, alongside with theoretical and empirical correlations. CFD shows clear benefits by allowing fast testing of new design concepts and configurations, by providing information even in cases where experimental activities are hard to accomplish, and by improving

Computational Fluid Dynamics Applied to Waste-to-Energy Processes
https://doi.org/10.1016/B978-0-12-817540-8.00001-7

© 2020 Elsevier Inc.
All rights reserved.

the understanding of the whole system leading to unexpected breakthroughs. Therefore, CFD simulation is a strategic asset to use in a large majority of the engineering processes [6].

In the particular case of WtE systems, the use of CFD would require significant efforts but not necessarily new technology breakthroughs. The next lines put in evidence some relevant examples. Efficient feeding strategies are possible by testing and evaluating different hydrodynamic features to prevent operational failures and reduce unnecessary procedures [7]. Boiler efficiency in waste plants is just about 30%, allowing a great margin for improvement [8]. The easy change of relevant parameters contributes to a thorough understanding of how they correlate with each other and how the engineer can optimize the full process, providing to the customer more effective and cheaper solutions [3]. The use of virtual reactors that differ only slightly from experimental data will allow considerable saves and a quick response to the market demands. Industrial case studies show that the testing time can be reduced up to half a year [9], and simulations of new standardized WtE plants will contribute to pushing the costs down [8].

However, CFD is not a simple subject! The equations governing the behavior of fluids are the result of over a century of hard and intense work. To make matters worse, complex mathematical tools are required to handle fluid behavior peculiarities. Consequently, students and even professionals are often overwhelmed in their attempt to connect their mathematical knowledge to practical applications. Furthermore, and as in any other computer approach, the use of CFD comprises a set of disadvantages and limitations that the user should be aware of and conscientious to reduce their impact [10]. Any CFD solution relies upon physical models and their predictions can only be as accurate as the models on which they are based. The same line of reasoning applies to the boundary conditions because their accuracy is only as good as the initial conditions included in the numerical model. Finally, computer solutions always imply round-off (finite word size available on the computer) and truncation errors (numerical model approximation).

The best way to gain proficiency is to understand how the CFD workflow can be broken down into manageable pieces, allowing the user to integrate the several steps involved and following through the entire process from A to Z. There are eight basic steps to implement any CFD attempt: (1) Modeling goals definition; (2) Domain identification; (3) Solid model geometry; (4) Mesh generation; (5) Configure physics; (6) Solver settings; (7) Compute solution; and (8) Model revision and improvements.

One common mistake when performing the modeling process is to fail in some of these steps leading to time-consuming and unnecessary procedures. In some cases, the user can take advantage of analytical approaches and get good insights for simple geometries where intensive computation is not necessary. Before committing to the simulation procedure, the user must first attempt an overall strategy concerning what it is intended to achieve.

These suggestions and steps are common to any CFD software available in the free or paid market. The user choice could depend on a large set of reasons as the availability, type of application, previous knowledge, or required features. Some paid programs provide free versions but often with several limitations [11]. Since there are numerous software packages to choose from Refs. [12–17], each one with their own set of settings, the scope of this book will focus on ANSYS Fluent framework. ANSYS Fluent applies for a large range of problems, offers a free version, and has a great number of users. Irrespective of the selected package, most of the approaches and strategies to solve the problems in the next chapters are quite similar.

In order to make strides in developing a valuable strategy and before providing the necessary basis to gain proficiency in implementing solutions applied to WtE systems, this first chapter is devoted to understanding how CFD is advancing along the last decades and how the flow analysis evolves through real problems. The chapter proceeds with a real case and details how CFD could explore new approaches and solutions that are hard to accomplish with bad decisions. Then, a review of WtE systems provides details of the different technologies and concludes how CFD helps their applicability and operational suitability. The remaining chapters present real cases of WtE systems by working through different decisions and providing computer screenshots, step-by-step guidelines, and some customization capabilities, allowing readers to move beyond button pushing and start thinking as professionals.

2. History of fluid mechanics

The first contribution in the fluid dynamics field dates as far back as III century BC when Archimedes developed the foundations of the fluid mechanics branch of hydrostatics [18]. He determined that the upward buoyant force exerted on a body immersed in a fluid has the same value as the weight of the fluid that the body displaces. This analysis drove the principles of flotation for ships.

A structured analysis in fluid dynamics lasted several centuries to come up, and it was only in XV–XVI centuries that Leonardo da Vinci provided outstanding contributions covering the movement of water, water surface, eddies, falling water, free jets, and numerous sets of other still unexplained phenomena. Relevant contributions were later provided for one of the most remarkable figures in science, Newton. His famous second law was applied to the interaction between fluids and bodies immersed in them allowing a quantitative analysis. He introduced the concept of Newtonian viscosity showing a linear relationship between stress and the rate of strain. Newton distinguished the fluids as in liquid or gaseous states, and the gaseous state comprises a set of noninteracting particles, which collide with the solid bodies immersed in it. Some of his conclusions on this topic were partially wrong and determined many doubts about the possibility of powered flights in the following centuries.

The XVIII century brought significant work regarding mathematical effort to describe the motion of fluids with special relevance for Bernoulli and Euler contributions. Bernoulli published important essays including quantitative analysis on how the force exerted over liquid conducts relates to the velocity of the liquid inside and to gravity. He also derived the famous Bernoulli's equation. Even more relevant in this century was the work of Euler for a better comprehension about the fluid behavior. The Euler equations described the conservation of momentum for an inviscid flow and the conservation of mass. These equations were the first good mathematical representation for the cases where density is constant (the fluid is incompressible) and for the cases where there is a relationship between pressure and the density (compressible gas flows not far away from ideal conditions) linked by the isentropic law for ideal gases. In these equations, the effect of viscous and heat conduction was neglected.

It would be unfair to not mention the work of d'Alembert who, some years before Euler's contribution, was close to writing the general equation for the conservation of the fluid's momentum. D'Alembert described effectively the conservation of the mass and the fluid acceleration, but he missed the relationship between acceleration and pressure. Names like Poisson, Lagrange, Poiseuille, John Rayleigh, Couette, and Laplace, among many others, also provided key contributions in this epoch [18].

In the XIX century, scientists were struggling with the inclusion of real flow effects into the Euler equations. Part of the problem was first overcome by Navier, who included the heat conduction effects that were not present in the Euler analysis. Some years later, in 1845, Stokes extended the Navier analysis deriving the equation motion of a viscous flow by incorporating the Newtonian viscous terms. This later contribution brought the Navier-Stokes equations to their final form and constitutes the basis of modern-day CFD. The resulting laws comprise a set of five differential equations (one for mass conservation, three for momentum conservation, and one for energy conservation) that with proper boundary conditions are able to predict the fluid velocity and corresponding pressure in a particular geometry. These equations involve a high degree of nonlinearity and complexity admitting only a few cases with analytical solution.

The implication of this is that for a long period and in the absence of strong computer capabilities, the experimental studies were the unique relevant part of fluid dynamics analysis. In fact, the starting point for computer work in the CFD field can only be traced back as far as the early 1940s with the pioneering work developed at Los Alamos National Laboratory with the first electronic computer [19]. Sadly, the genesis of these first attempts was intimately related to flow analysis for the H-bomb development and wartime technology. The first contributions were relatively simple and tried to solve the flow governing equations in the discrete space. However, and taking advantage of the advent of more powerful computers, Harlow, in 1957, at Los Alamos National Laboratory, proposed the famous particle-in-cell (PIC) method. This method suggests the use of a Lagrangian-Eulerian approach to describe the fluid motion. Several applications regarding shock interaction and supersonic wakes confirmed the method's effectiveness.

An important milestone is the work of Spalding and his team at Imperial College London in the 1960s and 1970s. They mainly focused on the combustion field but their findings were the foundations of numerical techniques such as the SIMPLE algorithm and the Eddy-Break-up combustion model. They also developed improved forms of the k-turbulence model [20].

During these decades, the description of incompressible flows was efficiently accomplished by incorporating implicit formulation of the equations and combining advanced solving procedures as the pressure/velocity solvers. This allowed the effective analysis of laminar and turbulent incompressible flows in different types of geometries.

The Imperial College in London was unable to proceed with the commercialization of their efforts in CFD codes and this, among other reasons, drove Spalding to create the CHAM Ltd. in 1974. One of the most challenging issues for CHAM Ltd. was to be able to handle and manage different CFD codes, and in 1980 they coupled all their codes into a single one called PHOENICS (Parabolic, Hyperbolic or Elliptic Numerical Integration Code Series). At nearly the same time, a group of scientists at Sheffield University released the first versions of FLUENT code and Engelman created a third company in this field called FIDAP (Fluid Dynamics Analysis Package). The products from all these companies were mainly based in Finite Volume and Finite Element methods using up to 10,000 cells and relying upon steady-state problems.

The 1990s brought meaningful advances in computer power and modeling and the typical CFD users, who were academics and developers, started to share this space with engineers working in R&D. The advent of introducing unstructured meshes into commercial Finite Volume codes revolutionized the ability to create simulations that are more realistic and new techniques to successfully cope with mesh motion. Incompressible and compressible flow solving schemes are now more effective by using new algorithms to speed up and increase the solver performance.

All these advances and the continuum powering of computers through the new millennium democratized the use of CFD beyond the pioneer disciplines of aerospace and nuclear engineering to the automotive, chemical, naval, bioengineering, and pharmaceutical industries. Fig. 1 summarizes the most relevant contributions in the Fluid Mechanics field through the centuries.

Despite the tremendous contributions of CFD for several industries, one should be aware that relevant simplifications still limit the full potential of this kind of analysis:

- Numerical approaches are always approximate models of reality implying important simplifications and assumptions. The user should be conscientious about the limitations of the physical model and assess the impact of such limitations.
- Because the computational domain where the flow analysis occurs sometimes applies to restricted zones, the use of boundary conditions should be wisely selected. Otherwise, serious errors can arise from improper conditions and jeopardize all the numerical simulations.

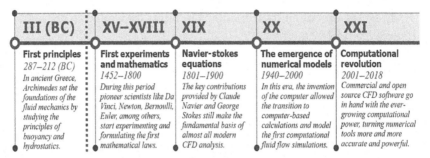

Fig. 1 Computational fluid dynamics throughout history.

- Discretization can highly affect the mesh convergence and solution accuracy. The user must consider which discretization formulation should be used in agreement with the geometry, mesh, and problem peculiarities.

3. CFD companies and resources

CFD is a priority tool for solving problems involving coupled phenomena such as the flow of fluids, chemical reactions, and heat, mass, and momentum transfer. The need for process understanding and improvement coupled with the right environment of exponential increase regarding the computational capabilities led to an exceptional proliferation of using CFD for targeting solutions that before required prohibitively expensive experimental work. The growing interest in this field and the demand for solutions by a large range of industries in different fields allowed the creation of several companies devoted to the development of highly effective solutions. Nowadays, CFD solutions are advantageously applied in a wide field of applications ranging from sports to chemical processing industries. Table 1 lists some of the most well-known and recognized companies and corresponding CFD products.

Most of these companies enlarged their skills presenting features that go far beyond the typical computation inherent to fluid dynamics and couple their service with solutions in robust design and optimization, uncertainty quantification, and other areas such as structures, electromagnetics, and systems, just to mention a few.

The examples given in Table 1 are mainly related to paid products and solutions. However, some packages like the ones provided by OpenFOAM and MFiX are free. Additionally, some student versions are also available (the ANSYS free student package is an excellent option still keeping a set of relevant features) but with some limitations regarding the number of cells or available models [11]. Anyway, these students' versions

Table 1 CFD companies and products.

CFD software	Vendor/ developer	Commercial	Main features
Fluent	ANSYS	Yes	Currently the most powerful CFD tool available in the market. Fluent provides a set of broadly well-validated physical models, capable of delivering fast and accurate results across a wide range of CFD, electronics, structural analysis and multiphysics problems.
CFX	ANSYS	Yes	Delivers reliable and accurate solutions over a wide range of CFD and multiphysics applications. Developed to deal mostly with rotating machinery problems, such as pumps, fans, compressors and turbines.
Barracuda	CPFD Software LLC.	Yes	A physics-based engineering software dealing exclusively with gas-particle fluidized reactors. Barracuda is capable of predicting all fluid, particulate-solid, thermal and chemically reacting behavior inside a fluidized reactor.
COMSOL	COMSOL Inc.	Yes	A cross-platform with finite element analysis, solver and multiphysics simulation software. It disposes of several modules dealing with electrical, structural, acoustics, fluid, heat, and chemical disciplines.
MFIX	NETL (Open Source)	No	General multiphase flow CFD code platform. It solves a generally accepted set of partial differential equations for the conservation of mass, momentum, species, and energy for multiple phases. MFIX has been broadly used for modeling bubbling, circulating fluidized beds, and spouted beds.
OpenFoam	OpenCFD Ltd. (Open Source)	No	The most widely used free/open source CFD software. OpenFoam has a broad range of features dealing with CFD, heat transfer, multiphase flow, molecular dynamics, solid mechanics and electromagnetics problems.
Star-CCM +	Siemens (former CD-adapco)	Yes	The second most dominant CFD software tool right after ANSYS Fluent. Capturing mostly the attention of aerospace, automotive and energy industries, Star-CCM + is a diverse platform dealing with CFD, heat transfer, chemical reaction, combustion, solid transport and acoustics.

allow users to practice and gain considerable insights and documentation is available, as well as webinars, where the users can learn the basic features of each software.

The following lines describe briefly some of these companies, the problems they usually solve, and additional information.

OpenFOAM [13] is perhaps the most famous open-source CFD software and its release dates back to the year 2004. The use of this software can be applied to a large range of problems involving complex fluid flows, chemical reactions, turbulence, heat transfer and in different fields as acoustics or electromagnetics. The software can be downloaded for Linux, Mac, and Windows environments and its release is every 6 months in June and December. OpenFOAM is a valuable option to solve the following problems, among others:
- thermal analysis, heating ventilation and air-conditioning;
- hydrodynamic features in vehicles;
- aerodynamics;
- combustion and standard multiphase problems;
- fluid-structure interactions.

Most often, the applications are ready to use and the user has access to all the coding features. In fact, the user can contribute to improving the code if he wants, in the spirit of collaborative research that characterizes this platform.

Jointly with OpenFOAM, the CFD code called Multiphase Flow with Interphase eXchanges (MFIX) is another valuable option free of charge. MFIX [14] is driven to solve multiphase problems and integrates NETL's effort to overcome the hurdles of simulating reacting multiphase systems. Contrarily to OpenFOAM, with a broader approach, MFIX devotes its efforts to problems characterized by multiphase nature.

MFIX provides advanced features to handle heterogeneous chemical reactions, interphase drag, polydispersity, particle attrition, particle agglomeration, and other complex phenomena. Any engineer should consider its use to model fluidized bed problems, gasification, and chemical looping combustion. NETL (National Energy Technology Laboratory) is one of the most important institutions in the world embracing fluidization research and with over 50 years of experience. Its facilities include a large set of experimental devices easing the numerical validation process. Similar to OpenFOAM, the user can download their releases in different environments.

ANSYS INC. [12] is perhaps the largest company in the world regarding simulation-driven product development with a vast portfolio in CFD and successful tools as FLUENT and CFX. Its approach relies upon parametric design optimization, simulation, virtual prototypes, and design analysis. It compiles the use of its solutions in a project management tool called the ANSYS workbench. The workbench serves to launch the individual software components and transfer data between them. It allows the transfer of data between geometry, mesh, solution setup, and postprocessing in only one environment. In the workbench, the user can observe how the models are built and the parametric analysis becomes easy because the user does not need to set up the process manually in each application. ANSYS INC. has been acquiring many other smaller companies to strengthen the range of its offer and provides valuable solutions in different areas such as systems, electromagnetics, acoustics, among others.

STAR-CCM+ from SIEMENS [15] is a simulation tool that provides a comprehensive set of physical models of any computer-aided engineering tool. Its integrated nature couples all the necessary tasks in only one environment, increasing the accuracy of its performance. In this interface, the user can use a broad range of validated models applicable to CFD, electromagnetics, aero-acoustics, heat transfer, and rheology. Some features facilitate to perform consistent and repeatable simulations, automatically updating the simulation pipeline to reflect any change in the geometry, mesh, boundary conditions, or solver setup.

These are just a few examples of codes and companies that allow the user to develop valuable solutions in a different range of problems. All of them rely upon a set of tasks that starts with problem identification and continues with the geometry, mesh, solution setup, and postprocessing steps. Relevant differences mainly come from the numerical method used to solve the Navier-Stokes equations: finite volume, finite element, hybrid, or both. In fact, the large majority of commercial codes still relies upon the Navier-Stokes equations implementation. However, it is fair to mention that some companies already use the Lattice Boltzmann approach.

In this method, one has to solve the kinetic equation for the particle distribution function, and the macroscopic variables such as velocity and pressure are obtained by evaluating the hydrodynamic moments of the particle distribution function [21].

Some advantages of the Lattice Boltzmann are:
- intrinsic linear scalability in parallel computing because the collisions are calculated locally;
- geometry complexity is easy to accomplish;
- interphase interaction modeling is effective because it is already related to the particle collisions.

On the other hand, this method is computationally expensive and is unsteady by nature. However, pseudo-states can provide stationary solutions. Relevant differences between using the Lattice Boltzmann (LB) or the Navier-Stokes (N-S) equations are as follows [21]:
- N-S equations are second-order partial differential equations while the LB method relies upon a set of first-order equations;
- in the LB method, the convection terms are linear while the N-S equations present nonlinear terms;
- N-S equations can take integral or differential forms while the LB method is characterized by a discretized kinetic equation;

- N–S equations are presented in the vector form making them independent of coordinate and grids while the LB method is intrinsically related to the lattice structure;
- the Navier-Stokes solver often uses iterative procedures to obtain a convergence while the LB method is explicit in form and the iteration process is not required;
- boundary conditions assume the form of particle distribution functions in the LB method;
- the interaction between phases can be easily incorporated in the LB method due to its kinetic nature.

One current company using commercial Lattice Boltzmann methods is the Exa Corporation from Dassault Systems [22].

Even with significant differences between the packages offered by the main companies, the approach to implement a CFD solution relies upon similar steps and stages. The next section intends to describe briefly how the user should organize the problem and implement the solution highlighting some of the most common pitfalls.

4. Simulation workflow

One typically misleading practice when implementing the simulation workflow is to consider the preprocessing stage as the first one. First of all, and most of all, the user should plan an overall strategy focused on what he intends to accomplish and before he decides on definitive actions, some questions must have a clear answer:

• What are the main results the user is looking for?
• How fast does the user need them?
• What degree of accuracy is required?

CFD is a powerful tool, but sometimes less complex approaches could be more efficient and save considerable time without it being necessary to compute time-consuming simulations. Additionally, simple analytical approaches could provide valuable insights into many problems and are usually a good starting point even for complex cases. This first step is crucial and can prevent a large set of future unnecessary procedures related to restarting all the analyses since the beginning.

After deciding which are the best options regarding the most suitable numerical approaches and the kind of results one is looking for, the user has now to identify where the numerical domain will begin and finish. Some decisions should be taken at this level:

- Is there information available about boundary conditions at the selected boundaries?
- Can the user extend the selected domain to a point where data are available?
- Is it possible to isolate the desired domain from a complex system?

These are just a bunch of questions that the user must consider before selecting the most proper domain for the numerical simulation. The user must understand that when he misses these first two steps and goes directly to the preprocessing stage, this means a strong leap of faith. The most likely advent for such a strategy will be a dramatic failure.

These first steps are the cornerstone to proceed toward the preprocessing stage. Now, the user should handle the geometry, mesh, boundary conditions, model physics, and solver settings issues. Perhaps this is the most challenging stage in our full numerical strategy. The preprocessing begins by defining a proper geometry to represent the domain of interest and then splitting it into small elements or cells. This last step, the mesh generation, will affect considerably the speed and accuracy of the upcoming simulation results.

The simulation proceeds with one more critical decision, which is the best model to describe the reality. When it comes to multiphase flows, the user must consider two main approaches:

- The Eulerian-Eulerian approach where both the gas and solid phases are treated as a continuum.
- The Eulerian-Lagrangian approach where the gas is treated as a continuum medium and the solid phase has its particles individually tracked.

Both approaches carry their own advantages and disadvantages. The Eulerian-Eulerian approach does not provide detailed data on the particles and the effects of particle size distribution are mainly neglected. However, it comes up with a lighter computational effort that makes the solution easier to converge and is not limited to an order of 2×10^5 particles as the typical Eulerian-Lagrangian methods. WtE processes always include chemical reactions and the user, once again, should evaluate how far he goes in the way the chemistry set is implemented:

- Is a detailed chemistry set the best solution?
- If using detailed chemistry schemes, which mathematical procedures should be implemented to make the converging procedure smoother?
- Is the user able to find reliable kinetic data to use as model inputs?
- Are the available forms of kinetic mechanisms enough to successfully describe the system or should the user develop his own code?

The use of detailed chemistry mechanisms is restricted to some case studies and always implies an entropy increase in the convergence process. The use of reliable kinetic data considering only relevant equations is usually an excellent starting point. The implementation of the physical and chemical models implies the selection of the most proper turbulence features (with the k-turbulence model being a first natural option), the materials and corresponding properties, the boundary conditions, and if the radiation needs to be considered or not considered.

This is the core of the simulation work, and the user should be familiar with all the relevant options and how they differ from each other. Each one of these options will be thoroughly discussed in the next chapters.

Having created a faithful representation of the domain, discretized it into a finite set of control volumes, and configured the physical, chemical, and boundary conditions based on the knowledge of the fluid application, it is now time to set up the solver settings.

The following is an outline of a general procedure on how to compute the solution:
- select the discretization scheme (and the pressure interpolation scheme when appropriate);
- choose the pressure–velocity coupling method (when appropriate);
- configure the underrelaxation factors;
- implement solver setting modifications if required;
- initialize the solution;
- enable the appropriate solution monitors;
- start calculation.

Default values are a good point of departure, but they do not usually ensure the success of the simulation. The solver comprises a large set of parameters, and their values often need to be adjusted to ensure a stable convergence and higher accuracy. This convergence is accomplished through an iterative procedure where the discretized equations are continuously solved until the changes in output variables from one iteration to the next are negligible. The overall accuracy of a given converged solution will mainly depend not only on the appropriateness and accuracy of the selected physical and chemical models and boundary conditions, and on the assumptions made during the mesh generation, but also on the way the user deals with numerical errors arising from a numerical simulation.

A common error found in any numerical implementation relies upon the user's attempt to achieve a stable convergence running all the model pieces in the first iteration. The user should start with a nonreacting flow and disable all the reactions, radiation equations, and fluid-particle interactions. The user should follow the numerical simulation adding the remaining submodels one by one only after a stable solution is found in each step. This procedure is of the utmost importance and will prevent wastage of a large set of hours without getting a stable solution.

When the user finally gets a feasible solution, the next step is to process it in meaningful and representative data. The users have a wide range of options when it comes to postprocessing options such as contour plots, vector, and streamlines, video screenings, creating planes and lines to study particular solution regions, appropriate graphical representations, and generate reports.

The process is not finished yet, and two disturbing thoughts should fill the user's mind:
- Is the converged solution the right one?
- How should the user proceed to validate the solution?

In fact, getting a converged solution does not mean that the said solution describes effectively the physical problem. There are two key systematic processes for confirming numerical results: the verification and the validation processes. The verification process

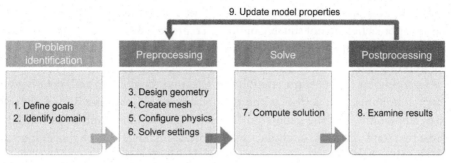

Fig. 2 How to implement a CFD solution.

is the procedure used to ensure that the program solves the equations correctly, while the validation process inspects the extent to which a model represents the reality by usually comparing experimental versus numerical data.

If the user is satisfied with the verification and validation procedures, and confident that the solution is adequate, he could plan some optimization actions. This can be done with complementary programs embedded in typical CFD software packages or by using the data as inputs in several statistical programs widely available. However, if the numerical data do not fit the experimental runs as expected, the user must reconfigure some options in the preprocessing stage and rerun the simulation. Fig. 2 depicts the stages, steps, and actions that the user should be aware of during a CFD simulation implementation.

The best way to show how a problem can be set up is by providing an example. The next section illustrates a simple approach to deal with a real problem using the suggested methodology described through this section.

5. Hydrodynamics in a fluidized bed gasifier—A case study

The example shown in this section intends to illustrate the capabilities of a CFD approach when applied to a real engineering problem. For the sake of simplicity, some simplifications are considered, and the following pages promote only a first approach but already with relevant insights. Many other features and considerations could be taken into account to turn the solution closer to reality. However, the focus here is to help the users to start their problem-solving with simple actions and only later with procedures that are more complex.

Previous guidelines suggest an overall plan with all the main goals clearly defined. In the present case study, the goal is to simulate the hydrodynamics of a biomass gasification process in a pilot-scale bubbling fluidized bed reactor. As main results, one intends to get a clear picture of the bed expansion response as a function of the inlet gas velocity and the biomass distribution patterns inside the reactor.

With these goals in mind and because one is looking for the distribution patterns, the 0D and 1D approaches are not suitable options. To prevent too complex approaches and because one does not intend to accomplish an extreme degree of accuracy, a 2D solution seems a good option.

It is now time to devote some attention to selecting the domain and identifying the boundary conditions. Fig. 3 depicts the pilot scale gasifier, the location of important boundary conditions, and the geometry. Further details can be found in Ref. [23].

These first steps allow us to advance with more confidence for the remaining stages and be sure that the CFD approach is proper to deal with the problem. Now, the user should start to prepare the preprocessing actions.

As mentioned before and to avoid complexity, the geometry will follow a 2D representation. It is important that the geometry building could faithfully depict the reactor dimensions (accurate height, base diameter, and inlet and outlet positions). Fig. 3 also provides the schematics of the reactor dimensions. The user must split the domain into cells creating the so-called mesh and should be aware of how far he goes in the number of cells. One should target a good balance between accuracy and expended time to get a good grid resolution. An improper grid may jeopardize the results and force the users to come back again at this point after wasting precious time in the upcoming stages. One can prevent such disturbing effects by proceeding with a mesh sensitivity analysis. The user should run simulations with four different meshes always doubling the number of cells. Then, it is important to select a response of interest and understand how the response changes. The discretization process starts with a coarser mesh, containing a

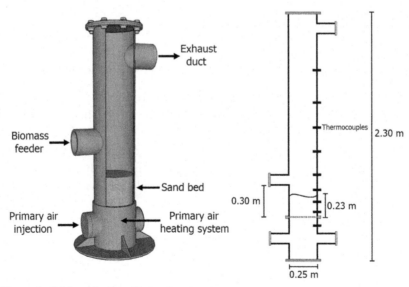

Fig. 3 Pilot scale fluidized bed gasifier and geometry.

higher element size, whose number of elements are then progressively duplicated, reaching a finer resolution with a smaller element size. A good criterion is to accept the mesh that presents similar results to the previous one. Fig. 4 shows the instantaneous contours comparison of the quartz sand volume fraction in four grids.

A brief analysis shows that the coarser mesh, with 12,502 elements, shows the greatest discrepancies as expected. Thus, this mesh is not capable of predicting the trends, outlining an unrealistic behavior, showing no true void inside the bubble space, along the centerline where only light green is seen, indicating the presence of bed material. Beyond this, it also gives misleading information regarding the bed height, which by comparison to the other meshes appears more expanded. The coarser mesh, 12,502 elements, and second meshes, 25,272 elements, show predominant bed material near the reactor's wall, evidencing an unrealistic behavior. The bed center region for the third, 50,544 elements, and finer meshes, 103,350 elements, is predominantly red dominated, meaning that the bed material tends to be in higher concentration in this region. Such behavior is consistent with the obtained quartz sand volume fraction results, in which sand particles were predominant at the middle of the bed height [23].

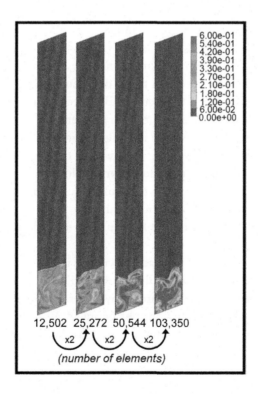

Fig. 4 Mesh analysis.

As expected, the finer mesh was the most predictable mesh (the fifth mesh does not show any relevant difference). However, it was 60% more time-consuming when compared to the third mesh, which, despite not being as accurate, was capable of showing good agreement and similar results when compared to the finer, unveiling consistency with the expected trends. In addition, being much less computationally expensive, thus allowing obtaining actionable conclusions in less time, makes the third mesh, with 50,544 elements, the reasoned solution considered for this work.

The user should not forget that any problem implies its own criteria. For instance, in the fluid dynamics of a fluidized bed gasifier, the cell size is related to the particle diameter. Some additional questions are not identified here but must be considered, and the user should be aware of them when solving a specific problem.

There are several approaches to model the characteristics of particulate flows [24, 25]. Table 2 depicts an overview of the main modeling approaches.

In this case, one will consider the Euler-Granular Model, where the dispersed solids are treated as an interpenetrating medium over the continuous fluid constituted by the gas. This model is quite easy to implement and is applicable both for dilute and dense particulate flows. The particle-particle interactions rely upon the Kinetic Theory of Granular flow, and different particle size distributions are possible by creating additional solid phases for different diameters. A more realistic approach considering the biomass hydrodynamics in a fluidized bed gasifier implies the use of a chemistry set with all the devolatilization, homogeneous, and heterogeneous reactions generally involved in this kind of phenomena. A deep analysis of the chemistry set implementation is provided in the following chapters. The implementation of an Euler-Granular Model is compatible and easy to link with all kind of reactions. By selecting the Euler-granular Model, the user must define a set of parameters: granular temperature model, granular viscosity, particle diameter, among others. As one is using solid and gas phases, their interaction must be taken into account by selecting an adequate drag model. Here, the user can consider some of the available built-in models or develop his own code and implement it through user-defined functions. In this particular case, the use of a customized code is crucial because built-in drag models are unable to capture a realistic fluidization behavior. For the sake of simplicity, the k-turbulence model describes the turbulence features.

The user can run the first set of simulations without including the chemical reactions to gain some previous knowledge about the system and then to evaluate the differences. Then, the user should notice that there are always several modeling options to treat any problem and the pros and cons of any must be balanced. With proper boundary conditions, one can consider system preheating or even gas flow changes. Relevant options considering boundary conditions, phase interaction, and chemical reactions can be found in Ref. [23].

Table 2 Characteristics of particulate flows modeling.

Models	Numerical approach	Particle-fluid interaction (F-F)	Particle-particle interaction (P-P)	Particle size distribution (PSD)
Discrete Phase Model (DPM)	Gas phase—Eulerian Solid phase—Lagrangian	Uses empirical models to determine subgrid particles.	Ignored.	PSD is easily included due to Lagrangian characterization.
Dense Discrete Phase Model-Kinetic Theory of Granular Flow (DDPM-KTGF)	Gas phase—Eulerian Solid phase—Lagrangian	Uses empirical models to determine subgrid particles.	Uses granular models to approximately determine the P-P interactions.	PSD is easily included due to Lagrangian characterization.
Dense Discrete Phase Model combined with the Discrete Element Method (DDPM-DEM)	Gas phase—Eulerian Solid phase—Lagrangian	Uses empirical models to determine subgrid particles.	P-P interactions are accurately determined.	PSD is easily included due to Lagrangian characterization.
Macroscopic Particle Model (MPM)	Gas phase—Eulerian Solid phase—Lagrangian	F-F interactions are determined as part of the solution. Each particle is assumed to span several computational cells.	P-P interactions are accurately determined.	PSD is easily included. Employs empirical model if particles turn smaller than mesh.
Euler-Granular Model	Gas phase—Eulerian Solid phase—Eulerian	Uses empirical models to determine subgrid particles.	Uses fluid properties like granular pressure, viscosity, drag, among others, to model the P-P interactions.	Employs PSD through population balance.

After setting up the model, the user should devote his attention to ensure the solution convergence. The default solver settings usually work reasonably for simple problems and are a good starting point to try the convergence. Some useful tips are:
- if running an isothermal simulation, deselect the equation energy;
- start the process with first-order discretization schemes;
- start the solution with conservative control settings;
- consider the gas operating density as zero;
- gradually increase the phases velocity;

- once the flow is established, increase the underrelaxation factors and try high order discretization schemes.

The user should define some criteria to consider the solution convergence and can follow those using monitors (pressure, temperature, or any other relevant parameter). The use of monitors allows the user to understand if the solver settings are good and how stable is the convergence process.

Unfortunately, the process does not finish when the solution converges because the criteria could be wrongly defined or even when well defined, the results could not fit the experimental data. After the convergence, the priority goes to the validation stage. Fig. 5 shows the experimental and numerical fluidization curves. Experimental and numerical data are in close agreement, indicating that our model seems to fit well the real physics of the problem. However, the user should extend this comparison to a large or at least relevant set of experimental data. This is a critical step, and the user must be extremely careful when dealing with real problems. Sometimes, one must be tempted to validate the models with a few points because experimental data are impossible to get or these are only available in different physical problems. This can lead to misleading conclusions!

When the user accepts the numerical solution, the next step is the postprocessing analysis. In this case, the goals are to determine the bed expansion response as a function of the inlet gas velocity and the biomass distribution patterns inside the reactor. Figs. 6 and 7 show both goals.

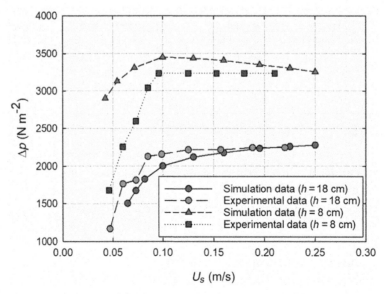

Fig. 5 Experimental and numerical fluidization curves for modeling validation purposes.

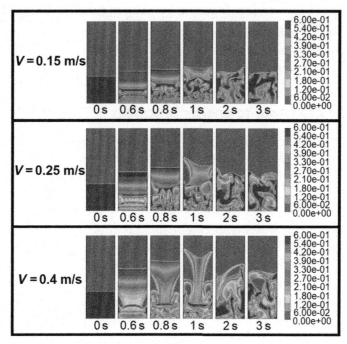

Fig. 6 Fluidized bed expansion response as a function of the inlet gas velocity.

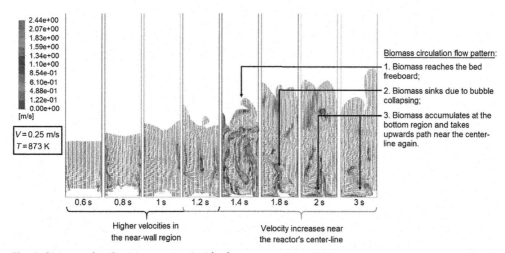

Fig. 7 Biomass distribution patterns inside the reactor.

With this case study, the author intends to show some modeling options in a real problem and focus the user's attention on several pitfalls and questions that will be raised through the simulation process. One will revert to similar case studies exploring several other issues as the chemistry set or the turbulence options.

6. CFD applied to waste-to-energy processes

Consumer society habits acquired throughout years of economic growth and urban development have led to an increased volume of generated Municipal Solid Waste (MSW). In 2015, the worldwide MSW generation was approximately 1300 million tons per year, with predictions dictating an annual growth rate of 4%–5.6% in developed countries and 2%–3% in underdeveloped ones [7]. By the year 2025, MSW should amount to an astonishing quantity of 2600 million tons per year, doubling its amount in a mere 10-year period. Such indicators come from assuming a current daily consumption of 1.2 kg of MSW per capita, increasing to an estimated 1.42 kg per capita by the year 2025 [7]. On the other hand, typical residues coming from plants or plant-based materials and agro-industrial companies (most often referred to as biomass) have been mentioned as a feasible solution, being looked at as valuable alternatives to fossil fuels. In the European Union alone, biomass dominates renewable energy consumption with a 68% stake and a 95.5% share for heat only applications and reached 8.4% of the total energy consumption in 2011 [7]. By the year 2050, the global biomass energy supply contribution is estimated to be around 160–270 EJ/year [7]. The biomass energy sector offers numerous routes for its energetic conversion, achievable by various thermochemical processes, namely combustion, pyrolysis, and gasification.

Thermal conversion of MSW in combination with biomass residues could play a key role in future energy production regarding not only environmental aspects but also due to the generation of new market segments for decentralized energy generation [26]. However, the highly heterogeneous nature of MSW implies significant variations in the product yield and quality. This impact can be reduced by blending this feedstock with others with more favorable characteristics and employing an effective pretreatment strategy that usually consists of size reduction, screening, sorting, and, in some cases, drying and/or pelletization to improve the handling characteristics and homogeneity of the material.

WtE conversion methods offer a promising solution not only to the ongoing energy problem but also in waste disposal, by reducing the volume of landfilled waste and increasing the recycling rates. Here, waste can be an important part of the equation as a huge and perpetual energy source due to its endless abundance generated by populations on a daily basis. Waste incineration was initially introduced as a potential solution, but the high associated costs, tight air pollution regulations, and ash disposal challenges made room for more viable alternatives [26]. Alternatives to incineration/combustion are

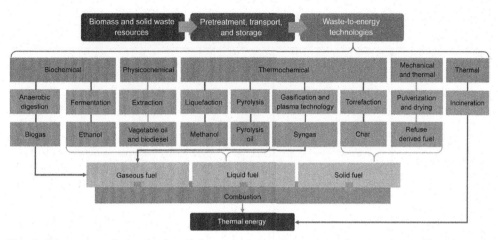

Fig. 8 Schematics of thermochemical processes and corresponding products. *(Based on M. Kaltschmitt, G. Reinhardt, Nachwachsende Energieträger, Grundlagen, Verfahren, ökologische Bilanzierung, Vieweg Verlagsgesellschaft, 1997.)*

gasification and pyrolysis, distinguishing from their ability to recover the chemical value from waste, rather than only its energetic value. Fig. 8 depicts the different thermochemical processes and corresponding products.

Gasification technology has been attracting researchers around the world due to its capability to use different feedstocks for different purposes, to obtain high efficiencies when compared to combustion processes, and to meet high goals regarding the pollutant emissions [27]. Gasification is also presented as a versatile technology able to be attractive for power generation on a small scale and allowing lower operating temperatures, which reduces the occurrence of serious operating problems such as alkali volatilization, fouling, slagging, heavy metal volatilization, and bed agglomeration (for fluidized bed reactors) [27]. Despite such advantages, gasification technology still faces some major challenges to target successful market penetration, being urgent to obtain a stable and cheaper syngas production able to meet higher environmental standards combined with high energy efficiencies.

Pyrolysis is thermal degradation either in the complete absence of an oxidizing agent or with such a limited supply that gasification does not occur to an appreciable extent; the latter may be described as partial gasification and is used to provide the thermal energy required for pyrolysis at the expense of product yields. It can be used for all types of solid products and can be more easily adapted to changes in feedstock composition than the alternative approaches. Table 3 lists the relevant features of incineration/combustion, gasification, and pyrolysis. Additional information is elsewhere [28].

The combustion/incineration process maturity leads to the use of this technology more often when compared with the other methods. Anyway, there are still gaps common to all

Table 3 Relevant features of thermochemical processes.

Process features	Combustion	Pyrolysis	Gasification
Operating temperature	800–1450°C	250–900°C	500–900°C with air; 1000–1600°C with other gasifying agents
Operating pressure	Usually atmospheric	Higher than or atmospheric	Usually atmospheric
Atmospheric medium	Air	None	Several possible gasification agents: air, pure O_2, O_2 enriched air, air-CO_2 mixtures, and steam
Stoichiometric ratio	Excess of air (>1)	Normally no air (0)	Substoichiometric air (<1)
Process outputs:			
Primary products	Heat	Char, liquid, and syngas	Syngas
Produced gases	CO_2, H_2O, O_2, and N_2	CO, H_2, H_2O, N_2, and other hydrocarbons	CO, CO_2, H_2, H_2O, CH_4, and N_2
Produced solid/ liquid phase	Ash and slag (*solid phase*)	Ash and coke (*solid phase*); Pyrolysis oil and H_2O (*liquid phase*)	Ash and slag (*solid phase*)
Pollutants	SO_2, NO_x, HCL, dioxins (PCDD/F), and particulates	H_2S, HCl, NH_3, HCN, tars, and particulates	H_2S, HCl, COS, NH_3, HCN, tars, alkali, and particulates

technologies that require further research. A few limitations that researchers must deal with are to:

- get reliable predictions regarding the effect of scaling different setups from the laboratory to semiindustrial or industrial dimensions;
- obtain reliable experimental data on large scales and with statistical effective treatments;
- drive the use of residues blends to produce high-quality products, by means of experimental and numerical approaches tied by validation and uncertainty quantification techniques;
- develop new catalytic materials able to promote efficient reactions without generating undesired by-products;
- optimize process configurations in order to minimize heat losses and improve the energy recovery efficiency;
- predict accurate pollutant emissions through a better understanding of the underpinning mechanisms behind the NO_x and other harmful types of emissions generated in these kinds of processes.

CFD approaches are relevant to minimize the impact of all these major issues. The next lines present a discussion on the topic.

There is a little prior work assessing the feasibility of gasification and pyrolysis runs using single and blended residues when it comes to large-scale experimentation. Experimental runs on a large scales are expensive and have inherent logistical issues, such as the use of efficient numerical models being highly required. Scale-up is not an exact science, being too complicated to use information collected in the laboratory to design a commercial reactor that can be tens or even hundreds of times larger. One of the relevant problems is that hydrodynamics phenomena in a laboratory scale fluidized bed are not the same as on large scales, exacerbating the situation that particles of different sizes do not scale up in the same way [7]. In this particular case, one can use a CFD model to compare the performance between different scaled reactors and get deep insights regarding bubble distribution along the bed height, residues velocity vectors profiles, average pressure drop curves slope, among other parameters. Fully model approaches allow getting temperature and species profiles. This is crucial for design purposes and allows preventing time-consuming and expensive experimental procedures.

The literature often uses one-factor-at-a-time (OFAT) procedures [3]. This means that important correlations between the relevant parameters are neglected. Design of Experiments (DoE) is a technique that allows for parameter interactions with a fewer number of runs. DoE is also an excellent tool to determine the optimal operating conditions to generate any particular product as well as the conditions to achieve substantial energy savings. CFD programs now integrate built-in DoE designs and optimization algorithms to carry out the full prediction and optimization work. For instance, it is now possible to simulate the product composition at the outlet of any reactor under a different set of operating conditions and target the maximum or minimum value of single or multiple responses of interest. This is a massive contribution to the engineering work. However, some of these points might set the process on a sharp peak of the response. Under these circumstances, one should select the operating conditions more robust to variation transmitted from input factors. These conditions can be found applying mathematical tools such as Propagation of error (POE) that are many times included in optimization toolboxes. To pursue this kind of analysis, it is necessary to get the standard deviation for the input factors, which can be obtained from experimental runs or by considering historical data. The way researchers are dealing with validation and uncertainty quantification techniques is now one active field in the use of CFD approaches applied to WtE processes.

Pollutant emissions are on the agenda of any researcher dealing with thermochemical processes. For instance, the emission of NO_x is of high interest to be reasonably predicted. The CFD approach implies that the NO_x transport equations are solved based on a previous flow field and combustion setup. This means that the NO_x predictions are dependent on how accurate the combustion solution is. Once again, this reminds us that the expected results are only as accurate as the input data and the developed simulation and selected physical models are. The state-of-the-art indicates that under most circumstances, the NO_x trends can be positively caught up, but the absolute values cannot be pinpointed.

Traditional biomass conversion processes still have only reasonable thermal efficiencies. Therefore, there is room for improvement of the use of biomass resources, which can lead to more sustainable and efficient solutions. The use of a CFD approach is perhaps the wiser way to extract meaningful information and proceed to target optimized and robust solutions by identifying the critical variable affecting the process as well as the objectives and the deliverables of any combustion, gasification, or pyrolysis process. Beyond the typical optimization of operating conditions, a CFD approach allows the testing of several configurations (catalytic reactors inside the main reactor, in situ catalytic measures, and cleaning actions) that would otherwise be prohibitive. Recently, CFD is being used to drive energy efficiency improvements in current setups by hindering heat squander and making the process more economic.

CFD packages have been experiencing a set of improvements in the last years by embedding powerful turnkey models. Despite such improvements, many of the included models present limitations and the user needs to introduce some customization actions to be close to real data. Customization is crucial, and this is always possible through the implementation of user-defined functions. For example, the hydrodynamics models normally included in CFD packages fail to predict real drag conditions. The user can develop a routine to adjust the coefficients from preincluded polynomial equations to enhance the standard drag model and heat transfer phenomena features within the applied mathematical model. The drag model and heat transfer adjustments are fulfilled by optimizing the default drag coefficients and gas–solid thermal conductivity parameters within their governing equations, allowing the customization of these general correlations to fit the particular modeling needs and to better agree with the experimental setup, regarding boundary conditions and material properties. Thus, the tuned routines provide a more accurate and predictable fitting procedure able to generate a more realistic behavior in different scaled reactors [23].

CFD advances are driving the momentum toward improved and more effective solutions regarding WtE systems. Most of the energy companies understand the use of CFD solutions as a valuable asset to make proper decisions in a very challenging and competitive environment as the energy market. The next chapters intend to be a hands-on approach guide on several real WtE problems, easing the user's proficiency on this topic.

7. Conclusions

The purpose of this chapter was to set the stage for learning about CFD solutions implementation regarding WtE systems and to give a brief overview of Fluid Dynamics history and WtE technology.

- Substantial advances in computer power and modeling in the last five decades made room for the development of CFD solutions in a broad range of industries.
- CFD companies are thriving with a steep demand not only from the traditional industries but also from emergent fields as electromagnetics, acoustics, or sports.
- The implementation of a CFD solution should follow a standard approach: (1) modeling goals definition; (2) domain identification; (3) solid model geometry; (4) mesh generation; (5) configure physics; (6) Solver settings; (7) compute solution; and (8) model revision and improvements.
- The use of a CFD approach meets the challenges that come up from the WtE systems allowing great technical improvements, more efficient designs and configurations, considerable energy and economic savings, and being an expeditious way to accelerate the process and prototype development.

References

[1] S. Yi, Y. Jang, A.K. An, Potential for energy recovery and greenhouse gas reduction through waste-to-energy technologies, J. Clean. Prod. 176 (2018) 503–511.

[2] P. Breeze, The economics of energy from waste, in: P. Breeze (Ed.), Energy From Waste, Academic Press, 2018, pp. 83–87 (Chapter 9).

[3] V. Silva, N. Couto, D. Eusébio, A. Rouboa, P. Brito, J. Cardoso, M. Trninić, Multi-stage optimization in a pilot scale gasification plant, Int. J. Hydrog. Energy 42 (2017) 23878.

[4] N. Couto, V.B. Silva, C. Bispo, A. Rouboa, From laboratorial to pilot fluidized bed reactors: analysis of the scale-up phenomenon, Energy Convers. Manag. 119 (2016) 177–186.

[5] N. Couto, V.B. Silva, E. Monteiro, S. Teixeira, R. Chacartegui, K. Bouziane, P. Brito, A. Rouboa, Numerical and experimental analysis of municipal solid wastes gasification process, Appl. Therm. Eng. 78 (2015) 185–195.

[6] J. Tu, G.H. Yeoh, C. Liu, Some applications of CFD with examples, in: J. Tu, G.H. Yeoh, C. Liu (Eds.), Computational Fluid Dynamics A Practical Approach, third ed., Butterworth-Heinemann, 2018, pp. 291–367 (Chapter 8).

[7] J. Cardoso, V.B. Silva, D. Eusébio, P. Brito, Hydrodynamics modelling of municipal solid waste residues in a pilot scale fluidized bed reactor, Energies 10 (2017) 1773.

[8] A. Granskog, P. Kåll, Bioenergy Threat or Opportunity? McKinsey on Paper Number 2,2012.

[9] E. Gardner, Can Computational Fluid Dynamics Improve the Food Processing Industry? https://www.foodprocessing-technology.com/features/featurecan-computational-fluid-dynamics-improve-the-food-processing-industry-5860298/, 2017. (Accessed 26 June 2018).

[10] F. Tabet, I. Gökalp, Review on CFD based models for co-firing coal and biomass, Renew. Sust. Energ. Rev. 51 (2015) 1101–1114.

[11] ANSYS, ANSYS Free Student Product Downloads, https://www.ansys.com/academic/free-student-products, 2018. (Accessed 26 June 2018).

[12] ANSYS, https://www.ansys.com/, 2018. (Accessed 26 June 2018).

[13] OpenFoam, The Open Source CFD Toolbox, https://www.openfoam.com/, 2018 (Accessed 26 June 2018).

[14] MFIX, https://mfix.netl.doe.gov/, 2018. (Accessed 26 June 2018).

[15] Siemens, Star-CCM+, https://mdx.plm.automation.siemens.com/star-ccm-plus, 2018. (Accessed 26 June 2018).

[16] Computational Particle Fluid Dynamics, Barracuda, http://cpfd-software.com/barracuda-vr-solutions/barracuda-basic, 2018 (Accessed 26 June 2018).

[17] Numeca International, https://www.numeca.com/, 2016. (Accessed 26 June 2018).

[18] K.A. Rosentrater, R. Balamuralikrishna, Essential highlights of the history of fluid mechanics, in: Proceedings of the 2005 ASEE Annual Conference and Exposition, 2005.

[19] J.S. Shang, Three decades of accomplishments in computational fluid dynamics, Prog. Aerosp. Sci. 40 (2004) 173–197.

[20] E.H. Hirschel, E. Krause, 100 Volumes of 'Notes on Numerical Fluid Mechanics', 40 Years of Numerical Fluid Mechanics and Aerodynamics in Retrospect, first ed., Springer-Verlag, Berlin Heidelberg, 2009.

[21] D.A. Perumal, A.K. Dass, A review on the development of lattice Boltzmann computation of macro fluid flows and heat transfer, Alex. Eng. J. 54 (2015) 955–971.

[22] Dassault Systemes, The Technology Behind PowerFLOW: Exa's Lattice Boltzmann-based Physics, http://exa.com/en/company/exa-lattice-boltzmann-technology, 2018. (Accessed 26 June 2018).

[23] J. Cardoso, V.B. Silva, D. Eusébio, L. Tarelho, P. Brito, Improved numerical approaches to predict hydrodynamics in a pilot-scale bubbling fluidized bed biomass reactor: a numerical study with experimental validation, Energy Convers. Manag. 156 (2017) 53–67.

[24] W. Zhong, A. Yu, G. Zhou, J. Xie, H. Zhang, CFD simulation of dense particulate reaction system: approaches, recent advances and applications, Chem. Eng. Sci. 140 (2016) 16–43.

[25] P.W. Cleary, J.E. Hilton, M.D. Sinnott, Modelling of industrial particle and multiphase flows, Powder Technol. 314 (2017) 232–252.

[26] N. Couto, V.B. Silva, A. Rouboa, Assessment on steam gasification of municipal solid waste against biomass substrates, Energy Convers. Manag. 124 (2016) 92–103.

[27] V.B. Silva, A. Rouboa, Combining a 2-D multiphase CFD model with a response surface methodology to optimize the gasification of Portuguese biomasses, Energy Convers. Manag. 99 (2015) 28–40.

[28] M. Kaltschmitt, G. Reinhardt, Nachwachsende Energieträger, Grundlagen, Verfahren, ökologische Bilanzierung, Vieweg Verlagsgesellschaft, 1997.

CHAPTER 2

How to approach a real CFD problem—A decision-making process for gasification

1. Introduction

Climate change is one of the greatest environmental, social, and economic threats of our time. To bypass these issues, a global shift toward the broad and sustainable use of renewable energy sources could help to fulfill the energy demands while mitigating the environmental problems [24]. Biomass products carry great development potential since they can be easily stored and transported, and unlike other renewable energy sources, they can also be converted into biofuels thus increasing their applicability, contributing significantly to the energy independence of the region along with associated economic and environmental benefits [64, 72].

With applications going back as far as the 19th century, gasification processes provide higher efficiency, cleaner energy conversion, and require less back-end pollution control equipment than other conventional processes such as combustion systems [45, 66]. Moreover, gasification has received renewable and sustainable interest in the thermochemical conversion of biomass for being CO_2-neutral, for handing over a high potential process, for improving the security of supply, and for its capability to provide both power, chemicals, and fuels solutions [25]. The same applies to solid waste residues, where gasification serves as a feasible technology in dealing with the massive amounts of waste produced daily in municipalities around the globe [69]. However, despite its many advantages and wide increased research, the stage of development for biomass and waste gasification can best be characterized as a limited niche development [45]. So, to stimulate further financial incentives along with the research interest toward developing the commercial use of biomass and waste gasification there are several technology setbacks that must be addressed, namely the scaling up [13, 26], tar and CO_2 production [27, 29], gas cleaning, and its economic competitiveness increase [28, 45].

Researchers have been trying to address these issues mostly using laboratory–scale gasifiers due to lower operating costs [33], once the work carried out on industrial gasifiers is limited given the high costs of a gasification plant that often reaches tens of millions of euros [5]. This is a major concern since a set of scale-dependent parameters, such as hydrodynamics, heat transfer, and particle residence time, change drastically when

Computational Fluid Dynamics Applied to Waste-to-Energy Processes
https://doi.org/10.1016/B978-0-12-817540-8.00002-9
© 2020 Elsevier Inc.
All rights reserved.

moving from laboratory-scale to commercial-scale reactors. Meaning that the information gathered in laboratory-scale reactors can be of little to no help when building a commercial size reactor that can be tens or even hundreds of times larger [13].

Mathematical models are being employed to work around this exact problem. Models provide a simplified representation of the real world, contributing to a much better understanding of the physical and chemical mechanisms laying within the reactor without major investments or time-consuming experiments [61]. Researchers use different modeling approaches reckoning on the rate of complexity they are willing to endure. Equilibrium models are a trendy method, due to their simplicity and ability to practically describe the gasification process, providing a quick way to calculate the maximum yield of the desired product [61]. Nevertheless, equilibrium models do not consider hydrodynamics, transport process, or reaction kinetics into account, resulting in occasional lack of meaningful information. These setbacks led to the development of kinetic models, being much more accurate but also computationally expensive [52].

The ongoing exponential growth in computational power is leading to a gradual replacing of empirical or semiempirical models for computational fluid dynamics (CFD) to predict biomass and waste gasification. CFD models provide crucial insights into the flow field inside the reactor, leading to a better understanding and improved performance of the operation while indicating solutions to potential problems. However, given the extreme complexity of creating a realistic model, this application is still in a developing stage and more studies are required [74, 75].

The gasification process is characterized by a multiphase flow containing solid and gas phases, in which slagging gasifier liquids can also be present. The solid particles may hold a wide range of sizes and shapes, especially when solid wastes are considered, while their organic components are being consumed as they pass through the reactor [11, 14]. The interplay between both phases is of utmost importance to model correctly since these exchange heat by convection, mass over the heterogeneous chemical reactions, and momentum due to the drag between gas and solid phase [73]. In most cases, this becomes an extremely complex process since the user must dispose of a thorough knowledge of all relevant phenomena involved (i.e., mass transfer rates, solid properties, heat transfer rates, reaction rates, the equation of state data and gas viscosities, among other features) which unfortunately are seldom available. Therefore, the modeling of these phenomena, beyond other required steps to create a realistic model capable of accurately predicting the gasification process, has proved to be a daunting process for most researchers.

Aiding researchers by providing the fundamental knowledge of the numerical tools available, is perhaps the easiest way to expedite the mainstream of this technology. With doing so, one can deepen both the fundamental knowledge of the phenomena occurring within the reactor, as well as optimize the process with more efficient operating conditions. A more efficient process will be more cost-effective, thus increasing the interest to commercialize it, while improving its environmental performance, which may lead governmental institutions to subsidize it or at least promote it. The literature may be extensive in presenting CFD studies regarding gasification processes [30–32, 55], however, lacks in providing a tutorial review carrying the starting point and relevant guidelines summarizing the fundamental steps to engage CFD simulations in biomass and solid waste gasification. Therefore, the purpose of this chapter is to fulfill this gap by delivering practical implementation guidelines for new researchers (and experienced ones) intending to undertake CFD analysis of gasification processes using the ANSYS Fluent framework. The reasons that led choosing ANSYS Fluent software as the scope of this implementation guide, resides in the fact that it is currently the most widely used and comprehensive CFD tool available, providing a robust set of well-validated physical models hence being broadly used by researchers to simulate gasification processes [4, 16, 32, 53, 70]. Thus, a thorough review of the work already developed in this field will be scrutinized, in which the whole CFD workflow will be broken down into manageable pieces, allowing the reader to follow through the entire process from A to Z. Also, all the main options will be addressed and the main research trends, alongside the most important results, will be discriminated. Tips and tricks for best practices for reaching the best possible results will be delivered throughout the paper.

2. Problem identification

One common mistake when performing the modeling process is to consider preprocessing as the first step. Before committing to the simulation procedure, the user must first endeavor an overall strategy concerning what is intended to achieve. Respecting gasification processes there are several questions that the user must answer before embracing a definitive strategy:

- What are the main results the user is looking for?
- How fast does the user need them?
- What degree of accuracy is required?

Depending on the matter at hand, there are different simulation approaches to solve a multiphase flow problem, ranging from simple analytical approaches to highly complex 3D CFD codes. Sometimes, it is challenging to determine the more appropriate approach from all the available tools. CFD is a powerful tool but its application should be restricted to cases where its implementation is justifiable since it can easily become much more computationally expensive and time-consuming than a simpler approach capable of reaching similar results.

Current literature divides the available tools for gasification into two main categories: mathematical and simulation models. The former comprises equilibrium, kinetic, and artificial neural network (ANN) models while the latter refers to CFD and flowsheet simulators [9]. Table 1 presents a brief pros and cons analysis on each.

Table 1 Brief pros and cons analysis of the available methodologies for gasification processes [9, 39, 68, 71].

	Pros	Cons
Equilibrium model	Simple, easy to implement and with quick convergence. Allows a practical description of the gasification processes with good approximation.	Poor process representation for lower operating temperatures. Inability to predict kinetics and hydrodynamic phenomena. Unfit for complex reactor designs.
Kinetic model	More accurate and detailed than equilibrium models. Capability to predict gas and temperature profiles inside the gasifier once it incorporates both reaction kinetics and reactor hydrodynamics.	Complex formulation and computationally expensive. Depends on reactions kinetics and type of gasifier.
ANN	Ability to be self-thought from sampled experimental data (machine learning). Able to represent complex nonlinear behaviors.	Slow convergence speed, less generalizing performance, arriving at a local minimum, and overfitting problems. Requires diligent work in implementing. Failure in case of limited data.
Flowsheet simulators	Helps to cut down on laboratory experiments and pilot plant runs. Can be used for risk-free analysis of various what-if scenarios.	Forces the user to think deeper about the problem at hands, in finding novel approaches to solve it, and to evaluate the assumptions closer. Process plants rarely operate entirely under steady-state conditions.
CFD	Highly precise. Very often offers faster resolutions than physical modeling. CFD model studies are generally 20%–40% less expensive than a comparable physical model effort.	It is generally run as a steady-state simulation. It carries a high degree of complexity; therefore, it requires incorporation of many engineering approximations, modeling shortcuts, and real-world variabilities.

CFD models are based on a set of governing equations for conservation of mass, momentum, energy, and species over a defined region of the reactor, capable of evaluating temperature, concentration, among numerous other parameters with a substantial precision rate [9]. Thus, CFD is broadly seen as one of the most appropriate and useful tools to deal with gasification processes, and for such reasons, it will be the core of this implementation guide. As computational power increases, researchers are allowed to create more complex models in a shorter time frame [42]. In fact, over the years the modeling developments seem to have a direct link to the exponential improvements to the processing power of modern computers. Beyond the ANSYS Fluent discussed here,

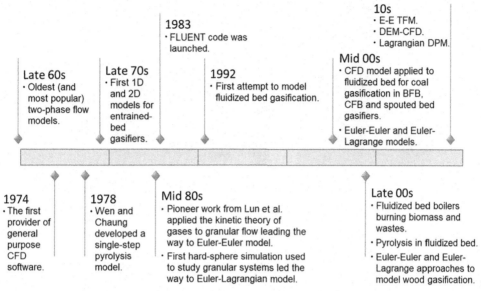

Fig. 1 Timeline concerning the main milestones in CFD gasification modeling.

there are more CFD tools available each one enclosing distinct features, thereby researchers must assess which software package is the most appropriate to solve the problem in hands. These can be segmented into commercial software such as ANSYS Fluent, ANSYS CFX, Star-CCM+, GASP, COMSOL, Barracuda, to mention a few, and also free open source solutions such as OpenFoam, MFIX, CFL3D, Typhon, OVERFLOW, and Wind-US. Fig. 1 gives a brief tail of some of the most important moments regarding CFD modeling of gasification processes. With the further advancements in computer and software technologies, researchers believe automatic design and optimization will become reality thus contributing to the use of CFD as a mature discipline and a powerful engineering tool in this field [49].

2.1 Identifying the domain

After carefully deciding on what type of results one wants to explore, and which is the most appropriate tool available, the user must focus on identifying the domain and isolate the intended section from the complete physical system. As a reference, Fig. 2 displays the 250 kW$_{th}$ gasification plant located at the Polytechnic Institute of Portalegre, Portugal. The schematic representation presents the main components that can usually be found in a gasification plant. Additional details concerning the experimental apparatus can be found in Ref. [73]. One must consider that these components can greatly differ depending on the design, gasifier type, feed system, along with other features.

As previously mentioned, most researchers tend to sort the complete system into their main components and focus only on the section intended for the study. Table 2 classifies

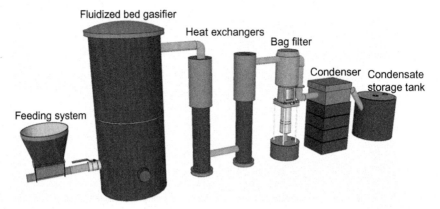

Fig. 2 Schematic representation depicting the main components of the 250 kW$_{th}$ pilot-scale gasification plant.

Table 2 Possible information and key observations for each domain [3, 62, 73].

Domain	Possible information	Remarks
Gasifier	Syngas composition Particle behavior Bed hydrodynamics Flow field Temperature distributions	The overwhelming majority of numerical studies focuses solely on the gasifier, such practice is understandable once it is the bottom line of the conversion processes. The purpose of choosing CFD analysis is to better understand the complex phenomena that occur inside the reactor, as well as optimizing the main operating parameters to ensure a stable operation.
Feed system	Velocity field Pressure gradients Particle behavior	A smaller segment of the gasifier studies also includes the feed system since it is directly linked to the performance of the gasifier itself.
Gas cooling system (*Heat exchangers*)	Thermodynamic analysis Thermal performance	Heat transfer rate and gas temperature are important features to capture and CFD analysis has been broadly employed to predict these quantities.
Gas cleaning system (*Filter and condenser*)	Catalyst production Tar decomposition Carbon deposition	Controlling syngas contaminants is one of the biggest deterrents to the commercial deployment of biomass gasification.
Entire plant	Feeding streams Unit operations Product stream	Studies concerning the entire gasification plant are mostly applied to flowsheet simulators like ASPEN Plus. Although this solution lacks to provide detailed results on particle behavior or aiding to understand the complex hydrodynamics within, it is capable of identifying all the major streams.

each domain and provides some of the possible information that the user may withdraw from each, alongside with some key observations.

This information can be extremely useful when determining the desired domain. As computational power and even more advanced models are developed, numerical models may give a helping hand in addressing other concerns mainly by upgrading reactor design and coupling the different systems.

3. Preprocessing

After defining the goals and choosing the appropriate tool to meet them, it is time to delve into the preprocessing. This step consists in defining a geometry to construe our domain of interest, dividing said geometry into elements, also known as the mesh generation step, and apply boundary conditions to the frontiers of that domain. The speed and accuracy of the results rely greatly on mesh quality and the ability to correctly prepare the model geometry.

3.1 Creating a solid model

After completing the preanalysis, creating geometry is the first major step in the problem specification. In the geometry step, one is affecting the mathematical model since it defines the domain in which the governing equations are defined. Moreover, boundary conditions are also defined at the edges of the domain.

During this stage, users may choose between two options, either using a preexisting computer-aided design (CAD) model or create one from scratch in ANSYS Design-Modeler. Starting with a preexisting geometry can save some time and effort, however, other challenges may arise such as how will the fluid region be extracted from a solid part. Also, trying to simplify or remove unnecessary features from an already defined solid can be tricky or time-consuming. When creating geometry from scratch, users are able to analyze the problem beforehand, allowing them to remove features they deem unnecessary that would complicate meshing such as fillets or bolt heads, which frequently add no crucial information and can be assumed to have little to no impact on the final solution.

For gasification processes, the above premise works just the same. Taking the example given in Fig. 2, one can easily create a simplified representation of the reactor faithful enough to conduct numerical experiments. ANSYS DesignModeler overview is depicted in Fig. 3a. Fig. 3b displays the simplified 2D and 3D version of our 250 kW_{th} pilot-scale fluidized bed reactor.

Regarding geometry modeling, there are some simplifications that one must consider in order to make the computation less expensive. Hence, most of the CFD models concerning fluidized bed reactors found in the literature apply 2D simulations rather than 3D

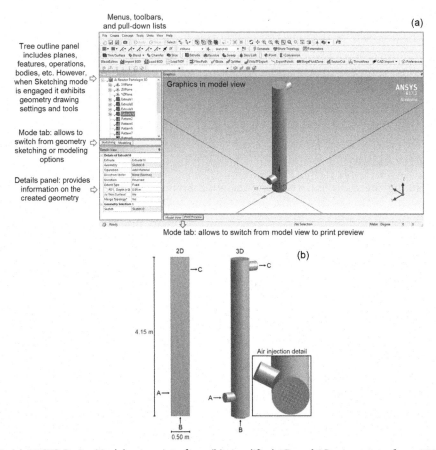

Fig. 3 (a) ANSYS DesignModeler user interface; (b) simplified 2D and 3D geometry of our 250 kW$_{th}$ pilot-scale fluidized bed reactor: (A) Biomass feeder; (B); air injection; and (C) exhaust duct.

simulations, given to its inherent computational cost [84]. Other authors go a step further and take advantage of design symmetry and use 2D axisymmetric problem setups [85]. In more "advanced" geometries some authors prefer to use 3D domains but only consider a small segment implying less computational time [49]. For the sake of minimizing computational cost authors try to maintain the geometry complexity as low as possible, yet, some authors put through a cost–benefit analysis, seeking to bring the geometry as close as possible to a real operating scenario in an attempt to achieve more realistic results. Thus, in fluidized beds, so to obtain fully developed turbulent velocity profiles at the air and/or biomass inlet, some authors include preinlet pipes for the oxidizer and/or substrate inlet (as detailed for the 3D geometry in Fig. 3b).

Table 3 Tips and tricks for solving geometry problems in ANSYS DesignModeler.

User problem	Solution
It can take many mouse clicks to achieve the desired function.	Use the middle-mouse-button multi options for easier navigation; Quick toggle single/box selection; Adopt DesignModeler "Hot Keys."
Requirements to visualize more easily areas of the geometry and understand issues quickly. Difficulty in finding intended details in a complex geometry.	Use the view options which can be accessed via the toolbar display controls and selection tools. Identify body interferences by enabling beta features; Reducing the display view to work effectively in smaller areas using Hide/Expand options.
Too much time spent building a large number of similar entities on a model.	The repair tool is designed to find and repair 2D and 3D holes; The pattern creation tool allows replicating an intended object without having to draw repeatedly from scratch, enabling to create several kinds of different repetitions.
Imported solids have gaps between them.	Several tools help the creation of contacts between parts in order to extract the fluid volume (i.e., thin surface; face delete; skin loft; fill by caps).

As previously established, one of the main goals of this chapter is to assist researchers with tips and tricks to enhance their capabilities when modeling complex systems not only in terms of reducing computational costs but also to achieve better results.

Table 3 displays some of the main problems researchers may find while creating the geometry, and what are the best solutions to solve them.

3.2 Designing and creating a mesh

Numerical methods require the geometry to be split into discrete cells, usually referred to as elements, this process is known as meshing or discretization. The ability of numerical methods to accurately predict results relies upon the mesh quality (i.e., elements distribution, elements quantity and shape, elements smooth transition). A mesh refinement process resides in solving the same case on progressively finer meshes. During this process, a typical good practice rule is to progressively duplicate the number of elements with each mesh created [14]. Also, for solid-gas CFD simulations, a ratio between the maximum grid size and the particle diameter should be considered depending on the case in hands [29, 44, 79]. Hereupon, "what level of refinement is needed to achieve accurate results?" The optimal answer is "just enough" to achieve a mesh independent solution, that is, mesh quality is sufficient so that it does not adversely impact the results. Moreover, the optimal mesh is the one that maximizes accuracy and also minimizes the solver

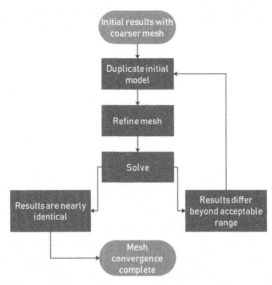

Fig. 4 Iterative process of mesh selecting flowchart.

run time. A common strategy to achieve this is adopting an iterative process as shown in Fig. 4.

This iterative process, also known as grid independence study, helps to avoid the use of an excessive number of elements while preserving the accuracy of the final solution. Fig. 5 displays a representation of this process applied to 2D and 3D geometries describing the reactor shown in Fig. 3b. The discretization process starts with a coarser mesh, containing higher element size, whose number of elements is then progressively duplicated reaching a finer resolution with smaller element size (as detailed within the circles).

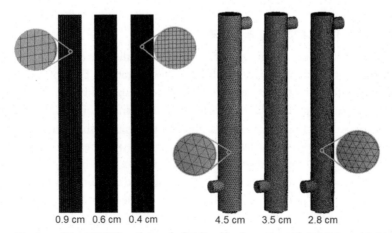

Fig. 5 Discretization of the 250 kW$_{th}$ pilot-scale fluidized bed reactor with 0.9, 0.6, and 0.4 cm element size for 2D geometry, and 4.5, 3.5, and 2.8 cm element size for 3D geometry.

Table 4 Main mesh quality parameters for the 2D and 3D meshes displayed.

	Quality parameters	Coarse	Medium	Fine
2D	Element size (cm)	0.9	0.6	0.4
	Number of elements	25,272	50,544	103,350
	Element quality	0.9281	0.9517	0.9981
	Aspect ratio	1.0686	1.0567	1.0010
	Skewness	$8.2534e^{-2}$	$5.6272e^{-2}$	$1.7071e^{-3}$
	Orthogonal quality	0.9895	0.9950	0.9999
3D	Element size (cm)	4.5	3.5	2.8
	Number of elements	84,854	167,750	340,208
	Element quality	0.8392	0.8455	0.8479
	Aspect ratio	1.8541	1.8223	1.8113
	Skewness	0.2212	0.2127	0.2093
	Orthogonal quality	0.8586	0.8640	0.8660

Table 4 displays some of the main mesh quality parameters for the presented 2D and 3D geometries. Features such as element quality, aspect ratio, skewness, and orthogonal quality are important indicators of the mesh quality. Element quality relates the ratio of the volume to the sum of the square of the edge lengths for 2D elements, or the square root of the cube of the sum of the square of the edge lengths for 3D elements, ranging from 0 to 1, in which higher values indicate higher element quality (0 standing for null or negative volume element, and 1 for perfect cube or square). Aspect ratio measures the stretching of a cell, and its acceptable range must be <100. Skewness provides the level of distortion of the existing elements from standard or normalized elements, hence skewness metrics must be kept as low as possible (0 for excellent, and 1 for unacceptable). Orthogonal quality is determined by vectoring from the center of an element to each of the adjacent elements, ranging from 0 to 1, where 0 claims the worst elements and 1 indicates high quality. Thus, regarding the presented quality parameters, the three meshes created for both geometries are within acceptance, with the quality standards improving as the meshes are made finer.

Mesh element shapes vary depending on the problem and the solver capabilities. For 2D geometries, elements may be set to triangular or quadrilateral shaped, while for 3D geometries these can be tetrahedral, hexahedral, pyramid, prism, polyhedral, or a combination of these. Recalling Fig. 5 mesh details, the presented 2D mesh comprises quadrilateral elements, while the 3D geometry uses mostly tetrahedral elements. Generally, quadrilateral and hexahedral meshes are often used to described simpler to moderate geometries, whereas triangular and combinations of tetrahedral, pyramid, prism, and polyhedral, are employed to describe complex geometries since these allow to create the mesh more easily over more intricate domains [7].

An efficient way to refine a particular mesh is by identifying where high gradients will be found. This can be accomplished in two ways: the first is by manually setting a particular region where high gradients are expected (i.e., near walls, inlets/outlets, wall boundaries, smaller features, and curved regions); the other possible route is applying a mesh adaption feature, allowing the automatic mesh refinement in regions the software sees fit without user interaction. However, in order to locally refine the mesh, the user must be able to recognize the areas where higher gradients are more likely to occur. ANSYS Meshing allows to speed up this process by implementing "Size Functions," these controls automatically refine the mesh in the areas that will typically have higher gradients. As a starting point, the meshes presented in Fig. 5 were built by applying the automatic default size function of "Proximity and Curvature," which showed to adequately capture the geometric features for this particular issue. After, the meshes were manually refined by implementing a conjunction of local size functions such as inflation layers, so to refine the mesh in the near wall region, predicting more accurately the flow gradients interactions near the walls within the reactor; and body sizing (face sizing for the 2D mesh), to control the mesh size of the entire geometry. Fig. 6 shows the ANSYS

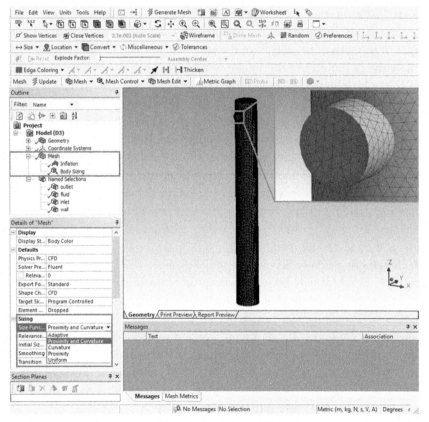

Fig. 6 ANSYS meshing outlining a simplified 3D mesh for the fluidized bed reactor with local meshing functions (inflation and body sizing) and automatic size function (proximity and curvature) applied.

Meshing outline with the applied mesh settings and details the inflation layers for the 3D mesh at the reactor's outlet.

When dealing with mesh quality one must consider the available computational resources in order to avoid outdoing them. Most researchers do not have access to computer clusters or supercomputers, also, most do not have access to complete packages of commercial software, instead, researchers adopt limited free academic licenses which carry certain restrictions such as the limited number of elements (512 K elements in fluid physics in ANSYS Academic license). With these free packages, users will dispose of limited mesh quality generation, and with poor computational resources, convergence may take too long to reach results.

Finally, this implementation guide is intentionally organized in a fashion to elucidate the reader which steps one must exert to properly model a gasification process, from top to bottom. However, for clarification purposes, in order to carry out with the hereby presented grid independence study, conventionally, the user must previously set, at least, some of the main features in the mathematical model, or the hydrodynamic model (whose implementation will only be deeply discussed throughout Section 5). This procedure allows the user to gather a set of initial results, suiting as means of comparison with the experimental data, allowing to assess beforehand if the settings made so far and the consequent obtained numerical results (even if still in a coarser stage) are within expectancy.

3.2.1 Mesh sensitivity analysis

To allow readers to have a deeper understanding of the relationship between mesh density, results, and computational cost, a mesh sensitivity analysis was conducted based on previous works developed by the research team [11, 13–15].

The simulations were carried out for 2D and 3D, with each configuration holding the three previously presented mesh resolutions, created with an increasing mesh refinement ratio of 2 and following the aforementioned maximum grid spacing rule of 10–12 times the particle diameter. All simulations concern gasification runs using eucalyptus wood as biomass (mean diameter of 5 mm) and quartz sand as bed material (mean diameter of 0.5 mm). Operating conditions were set at a superficial gas velocity of 0.25 m/s, time-averaged over a total of 3-s simulation time and operating temperature of 873 K. All geometries domains were created accordingly to the real dimensions of our 250 kW$_{th}$ fluidized bed reactor, 0.50 m width, 4.15 m height, and static bed height of 0.23 m, conceived to simulate the established experimental conditions as close as possible to a real operating scenario. Atmospheric air was delivered at the bottom (inlet) and the resulting syngas is removed throughout an opening placed at the top of the geometry (outlet). Additional information concerning the reactor configurations and simulations parameters can be found in Refs. [11, 13–15].

Fig. 7 shows the resulting time-averaged solids volume fraction contours for each of the three 2D and 3D mesh resolutions. Volume fraction contours are good indicators of

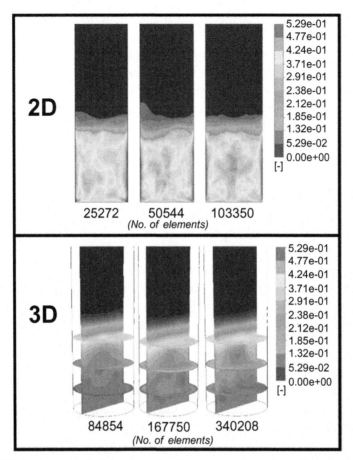

Fig. 7 2D and 3D time-averaged solids volume fraction contours for each studied mesh resolution. The 3D profiles are defined by a central plane cut at the middle of the reactor ($x = 0$) and by three additional axial planes ($z = 0.08$; $z = 0.18$; $z = 0.28$ m).

the mesh resolution effect, providing reliable information on the solids and void variation along the bed height. It is clear from these contours that by increasing the number of elements the clarity and definition of the solids or void occupied regions become more accurately defined. Regarding the 2D configuration, the finer meshes show clearer solids and void distribution (50,544 and 103,350 elements). On the other hand, the coarser mesh (25,272 elements) provides a roughest representation, depicting less solid presence at the centerline of the bed, area in which the following meshes point abundant solids presence. As for the 3D configuration, contours show that the coarser mesh (84,854 elements) is the greatest outlier of the set in portraying the solids distribution along the bed height, providing the worst resolution of the set. The 167,750 elements mesh presents a clearer solids volume fraction distribution compared to the previous mesh, by enhancing solids/void distribution along the bed height. The finer mesh with 340,208 elements shows the clearest solid formation and distribution, demonstrating the proper distribution of these quantities along the bed height.

Indeed, dissimilarities arise whenever mesh density is increased. In terms of mesh convergence, coarser meshes clearly tend to misinterpret the bed true behavior, pointing to erroneous assumptions and ambiguous results underlining the need for performing a mesh sensitivity study. Finer meshes are capable of providing a more accurate solution, nevertheless, computational time increases as a mesh is made finer, hence a satisfactory balance between accuracy and available computational resources must be executed. In this study, the finer meshes used to describe both configurations (103,350 elements for 2D, and 340,208 elements for 3D), were found to be substantially more over demanding to resolve, being about 60% more time consuming when compared to their corresponding previous meshes. Furthermore, when balancing the cost-benefit, the second meshes for both geometries not only revealed a good agreement regarding its aspect ratio but also were capable of addressing the same trends as the ones delivered by their respective third and finer meshes. Hereupon, by performing a mesh sensitivity analysis, it was found that the second meshes for both 2D and 3D geometries were able to achieve an accurate solution with a grid that is sufficiently dense to predict a proper flow hydrodynamics without overly demanding computational resources, with identical 2D/3D computational time ratio to the literature [85]. Additional information concerning mesh sensitivity analysis performed can be found elsewhere [11, 13–15].

To help users optimize their experience and ultimately obtain better results in quicker time, Table 5 displays some of the problems researchers find and what are the best solutions to solve them.

4. Setting up the solver

Having created a faithful representation of the domain and discretizing it into a finite set of control volumes, it is now required that the users configure both physics and solver values based on their knowledge of the fluid application. This setup stage is set within the ANSYS Fluent framework, requiring the general solver settings, the definition of physical and chemical mathematical models, material properties, phases creation and interactions, operating cell zone and boundary conditions.

When launching ANSYS Fluent a set of start-up options are made available. Here, the main listed options request specification on which type of geometry the user intends to solve, 2D or 3D; and the level of accuracy desired (enabling "Double Precision" is advised for enhanced accuracy). Along with these features, the user may also define the "Processing Options," allowing to set "Serial" (default option) or "Parallel" processing to solve the simulation. If the user disposes of a multicore machine, parallel processing can actively speedup the simulation, as it grants to select the number of cores (or

Table 5 Tips and tricks for solving meshing problems.

User problem	Solution
Manual meshing setup requires too much time.	When working with similar geometries, the mesh settings can be automated from an excel sheet thanks to Scripting.
Repetition of steps for CAD cleanup, meshing, setup, etc., for each design change.	Workbench can be used to templatize the work without having to know any scripting or programming language.
When geometry changes (especially with topological changes) entities selections are not always persistent.	Worksheet named selections can also be defined to make mesh setup persistent by location, size, type, or other criteria.
Difficult to pick or visualize areas of interest with a complex geometry.	Taking advantage of visualizing and picking entities inside ANSYS meshing.
It may take many mouse clicks to achieve the desired function.	Object generator tool will allow applying the same settings to several geometric entities.
Need to control the mesh size locally to achieve proper refinement.	Several tools are available to control the mesh size like bias, body of influence and virtual topology.
In some occasions, when meshing fails, it is difficult to detect the source of the problem.	The main source of errors usually comes from high coarseness, high skewness, large aspect ratios, large jumps in volume between adjacent cells, and inappropriate boundary layer selection. Identify and display highly distorted elements by using "Show problematic areas" and isolate problematic bodies.
The default behavior of inflation can be faulted when the geometry contains sharp angles.	A register key can be used to modify the smoothing.
By default, there is few information available about the meshing time.	Meshing performance report can be extracted from the Options menu.

threads) involved during the solving process. Unfortunately, the ANSYS student license is restricted up to a maximum of 16 cores.

Fig. 8 provides an overview of the ANSYS Fluent user interface. The "Tree" tab withholds the basic guide workflow on the various stages of how to perform a simulation in Fluent, starting in the read and mesh check process, up to the postprocess results. The most important setup setting steps for performing a gasification simulation will be briefly discussed. All remaining settings may be considered enabled by default. Given the complexity and broad suggestions to consider during the "Models" definition, this process step will be addressed in the following section of the mathematical model formulation. Moreover, assumptions on solution methods, simulation calculation, and postprocessing results will also be addressed further ahead.

In the setup general options, two types of solvers are available, "Pressure-Based" (default option) and "Density-Based." The pressure-based solver is used for most cases

Toolbar tabs: incorporate common settings and other important actions, all arranged based on typical stages in a simulation

Graphical display: the mesh is automatically loaded in the graphics window by default

Console window: controls the execution of the program and allows to input written commands

Detailed Tree tabs

Tree: contains the main simulation and setup settings, allows to define the models, set the solution methods and check the results

Task page: contains the various problem setting details from the selected tab on the Tree (on the left)

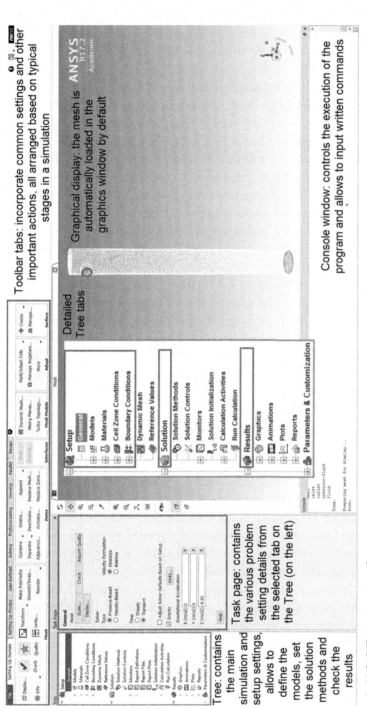

Fig. 8 ANSYS Fluent user interface.

(gasification processes included), handling problems with Mach number in the range of 0–2.3, once being more accurate for incompressible subsonic flows. On the other hand, the density-based solver is more accurate for supersonic flow applications with higher Mach number, such as to capture interacting shock waves. Regarding the time dependence, the flow characteristics can be specified as either steady state or transient. A flow is defined as steady state whenever the magnitude and direction of the flow are constant with time throughout the entire domain, while in transient the magnitude and direction of the flow change with time. Predicting in advance if the problem at hands calls for a steady state or a transient simulation may sometimes be a tricky step to grasp. In the event of a simulation running in steady state having difficulty in converging, even despite corrective actions by the user to retake the convergence such as mesh quality update or time step size adjustment, this might be an indicator of transient behavior. Thus, in case of reluctance, starting by performing a steady-state simulation is a good practice, once this method is less time demanding, allowing the user to save computational time as compared to a transient solution [7].

An important step in the setup of the model is defining the materials and their physical properties. Within the materials section, the user can edit (or create) the properties of any material from the ANSYS database. Here the user can input the relevant properties for the problem scope. ANSYS Fluent will automatically show the properties that need to be defined accordingly to the models previously selected by the user. For solid materials in gasification processes, density, specific heat, thermal conductivity, and viscosity must be defined. For each property, one may specify them as a constant; a linear or polynomial function; define it by a kinetic-theory, or even employ a user-defined function (UDF). UDF inclusion can be advantageous, allowing the user to customize the setup bringing the solution closer to its particular needs. In fact, ANSYS Fluent allows the user to customize a lot of its standard features by UDF inclusion. Such approach has been employed by the research team, in which a UDF was developed to enhance the material's absorption and thermal conductivity between gas and solid phases in fluidized bed processes [11, 15]. UDF text files once created ought to be saved in the same directory folder as the simulation file and can be implemented within the Fluent framework either by being interpreted or compiled in the "Parameters & Customization" tab. Interpreted UDF may be easier to set, however, compiled UDF run faster during the simulation calculation. In a fluidized bed simulation, it is common practice to create two separate material types, fluid and solid. The fluid region concerns the materials within the reactor's vessel namely, air, bed material, and/or biomass (if a binary mixture is considered), while the solid region relates the reactor's shell constituent material (aluminum is usually set by default).

The materials once defined one must set the phases and the interactions between them in the "Phases" dialogue box placed in the toolbars. The options contained here will vary regarding the type of multiphase model the user employs. In fluidized bed gasification processes, air is defined as the primary phase and the solid species as granular secondary phases. The granular phase properties section queries the user for the insertion of various parameters relating to the materials in question, such as the particles diameter, granular viscosity, packing limit, to mention a few. The user may add additional phases to describe additional solid species, for instance, to model biomass blends gasification, however, this will substantially increase the simulation time. Followed by this process step, it is time to establish the interactions properties. From the multiple active options, "Drag" and "Heat" settings lay among the most important features to define in fluidized bed systems. Here the user must be aware to correctly define the laws that rule the air-solid interactions and solid-solid interactions. Again, for a proper fitting of these parameters to the user's solution needs, one may opt to implement a UDF. Regarding the drag model, the author developed and included a UDF to improve drag in order to decrease deviations between experimental and numerical data. Additional details on how to define the various phases interactions in binary mixtures in fluidized bed gasification processes can be found here [14].

The cell zone conditions contain a selectable list of available cell zones from which the user can select the zone of interest (fluid region) and add the previously defined material types to their respective occupying cell zone. If applicable, other options may be defined such as porous resistance (for porous media simulation), heat sources, among other features. In this section, one must also set the parameters related to the operating conditions in the model, such as operating pressure (atmospheric pressure as the default value), temperature, density, and gravity. In fluidized bed simulations, the user must set the intended operating temperature, and most importantly, input the gravitational acceleration in the negative vertical direction to the geometry (as detailed in the "Task Page," Fig. 8). All remaining options may be set as default.

To finish the main actions to undertake in the setup stage, the user must now assign the boundary conditions to each previously designated zone and perform the required inputs for each boundary. Boundary conditions are a required and very important component of the mathematical model. These specify the boundary locations in the geometry, allowing to direct the motion of flow to enter and exit the solution domain. ANSYS Fluent provides various types of boundary conditions concerning the type of solution at hands and the physical models considered, so only the most commonly applied to gasification processes will be briefly discussed. Usually, gasification simulations carry four different main types of zones, inlet (flow input), interior-fluid, outlet (flow output), and wall; to each, a boundary condition will be assigned. The velocity-inlet boundary condition is usually employed to define the inlet, given to its suitability to describe incompressible flows (density of a fluid element remains unchanged along its path line). This condition applies a uniform velocity profile to the inlet (unless a UDF is used), in which the user defines the magnitude, direction, and temperature (if applicable) of the said flow. As for the outlet, the pressure-outlet boundary condition is generally applied, again due to its suitability to describe incompressible flows. Here, the user inputs the static gauge pressure of the environment into which the flow exits. Unmistakably, the interior-fluid and wall zones are appointed as the boundary conditions of interior and wall, respectively. No special considerations are required to set the interior-fluid, whereas, the wall boundary condition defines the region as a physical wall and allows the user to define parameters such as the walls temperature and the type of interactions the various phases will take with the wall (no slip or shear stress) [7].

Correctly setting a gasification model is one of the most complex processes in regards to current CFD state. To simulate the gasification process a multiphase (gas and solid phase) model describing the interactions between both phases must be employed. This has been achieved as exposed throughout the following section.

5. Mathematical model formulation

5.1 Modeling approaches for multiphase flows

When it comes to multiphase flows, the main distinction between CFD and the previously described methodologies is the level of detail the fluid mechanics is handled. Since the gas phase is usually described by a continuum approach, how one describes the solid phase dictates the approach:

i. if the solid phase is also treated as a continuum the approached is called Eulerian-Eulerian;

ii. if the solid phase has its particles individually tracked than the approached is called Eulerian-Lagrangian.

Until recently, modeling of gasification processes came down to two main methods: the Eulerian-Granular model (within the Eulerian-Eulerian approach) and the Discrete-Phase Model, or DPM (within the Eulerian-Lagrangian approach). Table 6 summarizes some of the main characteristics of the two.

Unfortunately, both of them carry their own set of limitations. The Euler-Granular model does not give any detailed information on the particles and the effects of particle size distribution are rarely considered [10]. Furthermore, the development of kinetic-

Table 6 Main characteristics of Euler-Granular and Discrete-Phase Model [10, 36, 40, 41].

Euler-Granular model	Discrete-Phase Model
Treats continuous fluid (primary phase) as well as dispersed solids (secondary phase) as interpenetrating continua.	The discrete phase is modeled by the Lagrangian model while the continuous phase is modeled by the Eulerian model.
Effects of particle-particle interactions are accounted based on the Kinetic Theory of Granular Flow (KTGF).	The discrete and the continuous phases are couple through sources terms in the governing equations.
Applicable from dilute to dense particulate flows. Particle size distribution can also be accounted by assigning a separate secondary phase for each particle diameter.	It is recommended to keep a volume fraction inferior to 10%. On the other hand, the mass loading can be rather large, in excess of 100%.
Compatible with species transport, homogeneous and heterogeneous reactions.	The DPM model accounts for the effect of turbulence on the particles trajectories.

theory closures is very challenging especially for the polydisperse particle group [40]. These setbacks are (mostly) solved with the DPM since it tracks every particle and calculates the interparticle collisions directly [36]. Thus, the DPM is fairly accurate but extremely computationally expensive, being currently limited to on the order of 2×10^5 particles and is often restricted to two-dimensional solutions without a fluid phase [41]. Hence, when modeling in pilot- or industrial-scale fluidized beds, where there are billions of particles to consider, one may quickly assess that the use of DPM is not a viable solution approach.

ANSYS Fluent framework incorporates an improved Lagrangian model known as Dense Discrete-Phase Model (DDPM), presenting enhanced grid independence behavior and easier particle size distributions implementation [20]. The DDPM is a Eulerian-Lagrangian parcel-based approach since the particulate phase is solved in a Lagrangian frame. Summarily, this approach extends the application range of the DPM by including dilute flows to dense particulate flows. Moreover, it considers the effect of volume fraction of the particle phase, and the particle-on-particle interactions while accounting for the fluid-particle coupling. Regarding particle-particle interactions, the DDPM can also be segmented into two main approaches the DDPM-Kinetic Theory of Granular Flow (KTGF) and the DDPM-Discrete Element Model (DEM) approach. The DDPM-KTGF approach is most suited for dilute to moderately dense particulate flows. Its main advantage is to allow for faster computations while predicting particle-particle collisions without full DEM. On the other hand, the DDPM-DEM approach is most suited for dense to near packing limit particulate flows [7]. As expected, this approach delivers higher accuracy at the expense of increased computational cost. Fig. 9 displays the entire framework of Lagrangian multiphase modeling currently available.

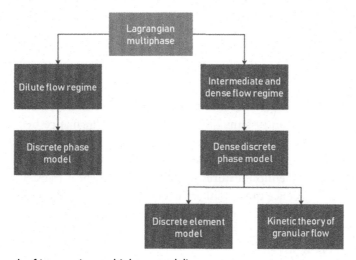

Fig. 9 Framework of Lagrangian multiphase modeling.

5.2 General governing equations

Gasification systems encompass a set of phenomena such as fluid flow, heat and mass transfer, and complex chemical reactions. CFD models solve these process phenomena by employing a set of governing mathematical equations, mostly based on conservation equations, namely, mass, momentum, and energy. Table 7 shows these fundamental governing equations. When employing a 2D or a 3D geometry domain, differences arise regarding their respective governing equations given that the mathematical model considers the boundary threads of the applied domain. Therefore, for simplification reasons, only the differences between 2D and 3D momentum governing equations (in polar and rectangular form) are demonstrated. As shown in 2D vs 3D momentum governing equations, x and y in the 2D expanded momentum equation refer to the r and y in the 3D cylindrical coordinate system, and, naturally, to x and y in the 3D Cartesian system. The additional third-dimensional terms are provided by θ (polar direction) in the 3D cylindrical coordinate system, and by z in the 3D Cartesian coordinate system.

CFD has the ability to implement these conservation laws over the discretized flow domain, computing the systematic mass, momentum, and energy changes that occur throughout a given simulation time as the fluid spans the boundaries of each discrete region (elements) [80]. Positively, the addition of a third dimension will not only reflect in equational differences but also provide dissimilar resolutions, in which the applied model will predict different operational quantities such as the reactor's hydrodynamics and syngas composition [84, 85]. Such is seen in Fig. 7, as the 2D and 3D meshes present different hydrodynamic behaviors, with distinct solids distributions along the bed height at the very same operating conditions (this subject will be detailed in Section 5.3). Thus, once more the user must balance between accuracy and computational demand when choosing to perform a 2D or a 3D CFD analysis. Additional information regarding the main differences between the 2D and 3D governing equations can be found in Ref. [84].

5.3 Hydrodynamic model

A proper understanding of gas and solids hydrodynamics is essential for the design and development of gasification systems since the reactors' performance heavily rely upon their hydrodynamic characteristics. Indeed, hydrodynamics phenomena carry a key role in describing gasification processes, whose parameters have been found to significantly influence physical and chemical features such as yield rates, heat transfer, mass transfer, chemical reactions, residence time, solid particles velocity, solids mixing and segregation, and erosion [11, 14]. CDF modeling valuably assists in understanding and analyzing these hydrodynamic parameters by providing an accurate prediction of the highly complex gas–solid mixture flow structure.

Table 7 General conservation model equations [73, 84].

2D governing equations

Mass:

$$\frac{\partial}{\partial t}\left(\alpha_q \rho_q\right) + \nabla \bullet \left(\alpha_q \rho_q v_q\right) = S_{pq}$$

Energy:

$$\frac{\partial}{\partial t}\left(\alpha_q \rho_q h_q\right) + \nabla \left(\alpha_q \rho_q \vec{v}_q h_q\right)$$

$$= -\alpha_q \frac{\partial(p_q)}{\partial t} + \overline{\tau}_q : \nabla \vec{v}_q - \nabla \vec{q}_q + S_q + \sum_{p=1}^{n}\left(\vec{Q}_{pq} + \dot{m}_{pq} h_{pq} - \dot{m}_{qp} h_{qp}\right)$$

Momentum:

$$\frac{\partial}{\partial t}\left(\alpha_q \rho_q v_q\right) + \nabla \bullet \left(\alpha_q \rho_q v_q v_q\right)$$

$$= -\alpha_q \nabla p_q + \alpha \rho_q g + \beta\left(v_q - v_p\right) + \nabla \bullet \alpha_q \tau_q + S_{pq} U_q$$

2D vs 3D momentum governing equations

Geometry

2D momentum equation in the expanded form:

$$\frac{\partial}{\partial t}\left(\alpha_q \rho_q v_q\right) + \frac{\partial}{\partial x}\left(\alpha_q \rho_q u_q v_q\right) + \frac{\partial}{\partial y}\left(\alpha_q \rho_q v_q v_q\right)$$

$$= -\frac{\partial P_q}{\partial y} + \frac{\partial}{\partial x}\left[\alpha_q \mu_q \left(\frac{\partial u_q}{\partial y} + \frac{\partial v_q}{\partial x}\right)\right] + \frac{\partial}{\partial y}\left(2\alpha_q \mu_q \frac{\partial v_q}{\partial y}\right)$$

$$-\frac{\partial}{\partial y}\left[\frac{2}{3}\alpha_q \mu_q \left(\frac{\partial u_q}{\partial x} + \frac{\partial v_q}{\partial y}\right)\right] + F_{pq}\left(v_q - v_p\right) - \alpha_q \rho_q g$$

3D momentum equation in cylindrical coordinate system domain (polar form):

$$\frac{\partial}{\partial t}\left(\alpha_q \rho_q v_q\right) + \frac{\partial}{\partial r}\left(\alpha_q \rho_q u_q v_q\right) + \frac{\alpha_q \rho_q u_q v_q}{r} + \frac{\partial}{\partial y}\left(\alpha_q \rho_q v_q v_q\right) + \frac{1}{r}\frac{\partial}{\partial \theta}\left(\alpha_q \rho_q v_q w_q\right)$$

$$= -\frac{\partial P_q}{\partial y} + \frac{\partial}{\partial r}\left[\alpha_q \mu_q \left(\frac{\partial u_q}{\partial y} + \frac{\partial v_q}{\partial r}\right)\right] + \frac{\alpha_q \mu_q}{r}\left(\frac{\partial u_q}{\partial y} + \frac{\partial v_q}{\partial r}\right) + \frac{\partial}{\partial y}\left(2\alpha_q \mu_q \frac{\partial v_q}{\partial y}\right)$$

$$+ \frac{1}{r}\frac{\partial}{\partial \theta}\left[\alpha_q \mu_q \left(\frac{1}{r}\frac{\partial v_q}{\partial \theta} + \frac{\partial w_q}{\partial y}\right)\right] - \frac{\partial}{\partial y}\left[\frac{2}{3}\alpha_q \mu_q \left(\frac{\partial u_q}{\partial r} + \frac{\partial v_q}{\partial y} + \frac{1}{r}\frac{\partial w_q}{\partial \theta} + \frac{u_q}{r}\right)\right]$$

$$+ F_{pq}\left(v_q - v_p\right) - \alpha_q \rho_q g$$

3D momentum equation in Cartesian coordinate system domain (rectangular form):

$$\frac{\partial}{\partial t}\left(\alpha_q \rho_q v_q\right) + \frac{\partial}{\partial x}\left(\alpha_q \rho_q u_q v_q\right) + \frac{\partial}{\partial y}\left(\alpha_q \rho_q v_q v_q\right) + \frac{\partial}{\partial z}\left(\alpha_q \rho_q v_q w_q\right)$$

$$= -\frac{\partial P_q}{\partial y} + \frac{\partial}{\partial x}\left[\alpha_q \mu_q \left(\frac{\partial u_q}{\partial y} + \frac{\partial v_q}{\partial x}\right)\right] + \frac{\partial}{\partial y}\left(2\alpha_q \mu_q \frac{\partial v_q}{\partial y}\right)$$

$$+ \frac{\partial}{\partial z}\left(\alpha_q \mu_q \left(\frac{\partial v_q}{\partial z} + \frac{\partial w_q}{\partial y}\right)\right) - \frac{\partial}{\partial y}\left[\frac{2}{3}\alpha_q \mu_q \left(\frac{\partial u_q}{\partial x} + \frac{\partial v_q}{\partial y} + \frac{\partial w_q}{\partial z}\right)\right]$$

$$+ F_{pq}\left(v_q - v_p\right) - \alpha_q \rho_q g$$

Engineering challenges such as scale-up effects in fluidized bed reactors can be effectively resolved by CFD analysis, minimizing experimental endeavor, avoiding drawbacks, and preventing the inappropriate waste of capital resources with building actual systems [13, 26]. To point this ability, Fig. 10a summarizes a previous analysis developed by the author's research group in which the hydrodynamics scale-up dependency was comparatively analyzed for two different-sized fluidized bed reactors, the already presented 250 kW_{th} and a smaller unit of 75 kW_{th}. Both units were simulated at the same experimental conditions, dimensionless superficial air velocity of U_0/U_{mf}= 2.5, operating temperature of 873 K, and total simulation time of 3 s. A dimensionless scaling method was used to correctly compare the hydrodynamic behavior. Main findings showed that due to its increased inner diameter, the 250 kW_{th} reactor presented superior near-wall velocity (less wall effect); increased bubble size, inducing higher gas rise velocity and consequently higher biomass particles velocity; and improved mixing behavior. On the other hand, the smaller 75 kW_{th} reactor presented higher centerline gas velocity (more pronounced wall effect due to smaller inner diameter), with solid particles behaving much more upwards oriented and less dispersive [13]. Therefore, scale-up does largely influences on the hydrodynamic phenomena.

When performing a 2D or a 3D gasification analysis, the presence (or nonexistence) of a third dimension will also affect the hydrodynamic flow behavior throughout the domain. To show this hydrodynamic model sensitivity to the applied domain, the previously selected 2D (50,544 elements) and 3D (167,750 elements) domains are comparatively addressed. Fig. 10b shows the 2D and 3D instantaneous quartz sand and eucalyptus wood biomass volume fraction contours, jointly with the 2D and 3D bed expansion at different superficial gas velocities (0.25, 0.40, and 0.60 m/s). Operating conditions were kept the same for both domains at a temperature of 873 K, initial superficial air velocity of 0.25 m/s, and total simulation time of 3 s. Overall, a proper solids distribution is delivered by both configurations, with the bottom bed sections containing mostly quartz sand (heavier), while the top bed sections enclosing essentially eucalyptus wood biomass (lighter). Yet, a sharper representation of the solids separation along the bed height is provided by the 3D configuration, while the 2D considers a fade-out behavior between the solids separation. This may be given to the more accurate and realistic behavior analysis broadly addressed to 3D simulations [35, 50]. The dimensionality effect is also quite visible in the bed expansion, with the 2D configuration tendentiously overestimating the bed expansion as compared to the 3D configuration, particularly at higher velocities of 0.6 m/s (around 30% more). Such behavior is justifiable as the wall area and the reactor volume ratio is larger for the 3D configuration, thus the overestimation provided by the 2D configuration deem reasonable [21].

Often, default submodels within the ANSYS Fluent framework do not provide the level of accuracy intended to fulfill the researcher's modeling needs. In fluidized bed

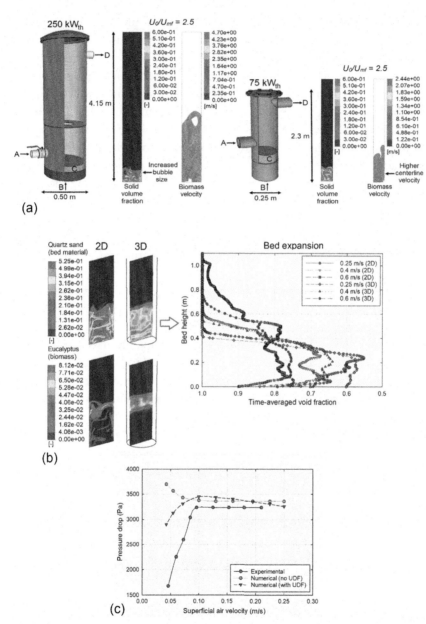

Fig. 10 Hydrodynamic model: (a) Comparative scale-up analysis concerning the solids volume fraction and biomass velocity vectors for two different pilot-scale fluidized bed reactors: (A) Biomass feeder; (B) air injection; (C) bed; and (D) exhaust duct; (b) 2D and 3D instantaneous quartz sand and eucalyptus wood volume fractions (3D representation at the central plane cut, $x = 0$), alongside with the bed expansion given by the time-averaged void fraction profiles at three different superficial gas velocities along bed height; (c) Experimental and numerical fluidization curves, including UDF improved numerical values.

reactors modeling, the fluidization velocity has significant influence over the hydrodynamic flow. Despite the rigorous mathematical modeling and rather accurate physical assumptions, the drag laws provided by Fluent, used to predict the fluidization conditions, generally require certain customization to particular modeling needs namely, geometry, boundary conditions, and material properties. Thus, ANSYS Fluent allows enhancing the hydrodynamic behavior during gasification simulations, allowing to couple simple program add-ins, UDF. Fig. 10c depicts the deviations between experimental (gathered from the previously presented 75 kW_{th} pilot-scale fluidized bed reactor) and the numerical fluidization curves, including the numerical improvements provided by the UDF. The implemented UDF acts over the drag model by tuning the default drag coefficients within the governing equations, these adjustments allow to alter the general correlations within the Fluent database to fit our intended modeling needs and to better agree with our experimental setup. Its application revealed a significant effect by providing a better agreement between the numerical and the experimental results gathered from the fluidized bed reactor [12, 14].

To accurately predict the fluidization behavior, the shape and general properties of particle size distributions, such as density and diameter, must be considered once the hydrodynamic flow will largely rely upon these quantities. Dealing with perfectly spherical particles carries no special concern, however, in reality, biomass particles are usually rather irregular. Modeling the interactions of millions of biomass and bed material particles in a pilot-scale reactor, detailing the particles irregular shapes by coupling structural submodels, is discouraging and computationally overdemanding process. Thereby, a set of simplifications can be applied to save both time and unnecessary effort. Commonly, equivalent spherical diameter approaches are used to represent these irregularly shaped biomass particles. These approaches employ some form of mean values for the different properties of the single particles, setting them either as spheres with the same surface area-to-volume ratio, spheres with the same length, same surface area, or with the same volume as the original irregular shaped particles [19].

Given the multivariate of hydrodynamic parameters and their effect on the gasification quality, further details regarding the hydrodynamic model implementation can be found in the following works by the author's research group [11, 13, 14].

5.3.1 Turbulent flow

The great majority of flows in industrial processes are of turbulent nature, and turbulence prediction is still one of the main challenges in classical physics. Turbulence plays an important role in the gasifiers flow phenomena, particularly at high Reynolds numbers (a dimensionless constant which relates kinetic energy and viscosity in fluid flow). Turbulent regimes are described by fluctuating velocity fields, having a critical influence on hydrodynamic quantities such as the motion of transported quantities, affecting over momentum, energy, species concentration and transport (enhancing mixing), heat transfer, drag, vorticity distribution, and swirl flow [16]. To calculate turbulent flows there are three major modeling approaches based on various simplifying starting assumptions:

- direct numerical simulation (DNS);
- large eddy simulation (LES);
- Reynolds averaged Navier-Stokes simulation (RANS).

The DNS approach numerically solves the full unsteady Navier-Stokes equations without requiring any modeling. However, since it details too much information it is excessively computationally expensive for nowadays computational capabilities, being restricted to simple turbulent flows hence unappropriated for complex industrial problems. In the LES approach, the motion of larger eddies is directly resolved in the calculation while the smaller eddies (smaller than the mesh) are modeled. This approach while less expensive than DNS, its use is still too computationally expensive for most practical applications. Finally, the RANS approach is the only modeling approach for steady-state simulation of turbulent flows, being currently the most widely used approach for industrial flows [80].

Within the RANS approach, there are more than a dozen turbulence models, each one with several auxiliary models and near-wall options. Excluding the simpler one-equation models (such as Spalart-Allmaras model), and the more advanced and accurate turbulence models (such as the five-equation Reynolds Stress model), the most common RANS models are the two-equation models in which the solution of two separate transport equations allows the turbulent velocity and length scales to be independently determined. Table 8 displays the five two-equation turbulence models available and their respective recommended usage. The models are numbered from the least (standard k-ε Model) to the most computational costly (SST k-ω Model) per iteration.

Unfortunately, no single turbulence model is universally accepted as being superior for all classes of problems. Also, none of the current existent turbulence models can reliably predict all turbulent flows with adequate accuracy [6]. Thus, choosing the appropriate turbulence model requires major concern and will rely on a set of considerations such as the physical nature of the flow, the established approach to solve a specific class of problem, the accuracy required, and the computational cost. The same process is applicable to modeling gasification processes. However, the bulk of studies published on the subject tend to employ the standard k-ε Model [16, 51, 87], mainly due to its robustness and reasonable accuracy for a wide range of flows, becoming the workhorse of practical engineering flow calculations since it was first proposed by Launder and Spalding [47].

Table 8 Two-equation turbulence models and their respective recommended usage [7].

Model	Recommended usage
1. Standard k-ε Model	Default k-ε Model, and the most widely used engineering turbulence model for industrial applications. Robust and fairly accurate, it is broadly employed despite its known limitations as only valid for fully turbulent flows, and its poor performance for complex flows, including harsh pressure gradient, separation, and strong streamline curvature. Suitable for exploring basic flow pattern and parametric studies. Appropriate for converging initial case before switching to other models.
2. RNG (Renormalization-group) k-ε Model	Improved accuracy and reliability for a wider class of flows than the standard k-ε Model. Recommended for complex shear flows involving rapid strain, moderate swirl, vortices, and locally transitional flows.
3. Realizable k-ε Model	Improved prediction for spreading rate of jets, superior ability to capture the mean flow of complex structures and for flows involving rotation, boundary layers under strong adverse pressure gradients, separation, and recirculation. Delivers improved accuracy and easier convergence than RNG.
4. Standard k-ω Model	Default k-ω Model. Most widely adopted in the aerospace and turbo-machinery communities. Improved performance for wall-bounded boundary layer, free shear, and low Reynolds number flows, however, requires mesh resolution near the wall. Not recommended from an industrial standpoint except if the user disposes of good boundary conditions for k and ω.
5. SST (Shear-Stress Transport) k-ω Model	Carries similar advantages than the standard k-ω Model. Recommended for high accuracy boundary layer simulations. Wall-bounded flow (i.e., blades, airfoils, compressors, turbines, etc.). Appropriate prediction of wall heat transfer and for mildly separated flows (without large separation).

5.4 Chemical reactions model

Chemical reactions are present in most industrial processes which turn the chemical reaction model into an extremely important asset to implement. Setting correctly a reacting flow in CFD can be an intimidating challenge since one will have to deal with a wide range of time and length scales, a large number of transport equations, and complex turbulence–chemistry interaction.

Turbulence has an extremely important role in reacting systems. Reacting flows can be segmented into fast chemistry and finite-rate chemistry depending on how the mixing timescale (τ_F) relates to the chemical timescale (τ_{chem}), also known as the Damköhler number (Da). Table 9 provides a quick review of the two methods.

Table 9 Brief review of fast versus slow chemistry [43].

Fast chemistry	Finite-rate chemistry
$Da \gg 1$.	$Da \sim 1$.
Chemical reactions limited by turbulent mixing.	Chemical reactions limited by chemistry and turbulence interactions.
Selection of turbulence closure model is important.	Turbulence/chemistry interactions are important.
	Selection of reaction mechanism is important.
Deals with chemical reactions associated with combustion in furnaces, boilers, gas turbines, gasifiers, incinerators, flares, etc.	Deals with chemical reactions associated with pollutants formation, ignition and extinction, chemical vapor deposition, nonequilibrium phenomenon, and air dissociation at hypersonic speed.

Gasification is a thermochemical conversion process that occurs in an oxygen-deficient environment to convert carbonaceous wastes into synthetic gas (also known as syngas) or chemical feedstock. A typical gasification process consists of a series of stages that overlap each other without a clear boundary between them. Fig. 11 shows a simplified form of modeling and understanding the multiple stages involved throughout the gasification process.

Carbonaceous materials such as biomass (or solid waste) particles enter the reactor at room temperature and the process commences by increasing the operating temperature until moisture is released from the feedstock (around 433 K). This drying stage process can be modeled by a common wet combustion model or by applying user-defined moisture release law [63]. The following stage, the devolatilization, occurs at temperatures up to 973 K. Here, the existing solid residues go through thermal cracking to produce light gases, tar, and char. The products from this thermal degradation react with each other and with the gasifying agent (which can be air, oxygen, steam, carbon dioxide, or a combination of these) to form the final gasification products [22, 25, 27]. Since devolatilization and gasification reactions steps are critical in the overall process, a detailed analysis will be addressed in the following sections.

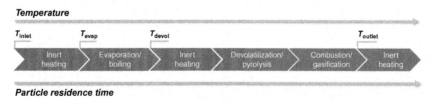

Fig. 11 Gasification process depicted by the solid particle residence time as a function of the temperature. *(Modified from ANSYS, 2013. Modeling Reacting Flows in ANSYS CFD, Part-IV: Modeling Solid/Liquid Combustion.)*

5.4.1 Devolatilization submodel

The devolatilization process occurs after initial drying. In this process, the feedstock is thermally decomposed into volatiles, char, and tar. To do so, chemical bonds within large organic molecules must break so that hydrocarbons subgroups can be released and evaporate to form volatile matter. This matter is usually segmented into permanent gases (H_2, CO, CO_2, CH_4, water vapor) and condensable gases (tar). Generally, CFD models deal with solid fuel particles by coupling them with fluid flow solutions via source terms acquired from Lagrangian single particle calculations. In gasification simulation, capturing the species concentration during the devolatilization process is essential, turning the devolatilization submodel in one of the most important to establish. These submodels operate by calculating the volatile gases released from the particles, commonly defined by global Arrhenius kinetics [59]. The devolatilization process is extremely problematical leading researchers to employ simplified approaches to handle it. Thus, the intricate devolatilization reactions are frequently combined in single global reactions (or in multiple parallel global reactions) under a single set of Arrhenius parameters, allowing to save computational effort while being suitable for large-scale industrial simulations. The more advanced devolatilization submodels allow estimating the mass loss of solid particles based on the kinetics of various chemical groups enclosed in the solid fuel [59].

Several devolatilization modeling approaches are available in the literature [18, 54, 77]. Most recently, Nakod [57] developed an advanced devolatilization submodel approach, known as a volatile breakup, employing a step-by-step conversion of the volatile matters from solid fuels into the gas-phase species concentrations. Fig. 12 outlines this step-by-step volatile breakup algorithm set to calculate the mass fractions of the resultant species. This approach is suitable for any sort of solid fuel as it conserves the mass of each element and

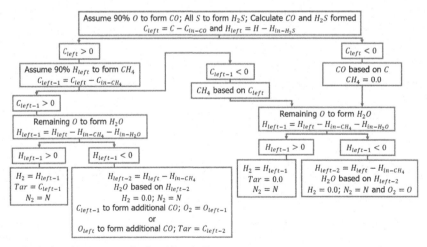

Fig. 12 Volatile breakup approach algorithm [57].

overall heat content during the conversion process. The following gas-phase volatile breakup reaction is used to convert the gaseous volatile into other gas-phase species:

$$Volatiles \rightarrow aCO + bH_2S + cCH_4 + dH_2O + eH_2 + fN_2 + gO_2 + hTAR \qquad (1)$$

Additional data on volatile breakup approach is available in Refs. [57, 58].

Comparatively to the work of Nakod [57], our research group adopts a simpler single-rate model composed of two pseudo-heterogeneous reactions (devolatilization and demoisturization) modeled in a single reaction rate in agreement with the Arrhenius law [73]. Table 10 summarizes the devolatilization submodel equations applied to biomass thermal decomposition and their respective Arrhenius reaction rates. This approach has main advantages that rely on moderate and reliable devolatilization rates with little computational demand. Further details on this devolatilization submodel implementation can be found in Ref. [73].

In order to deal with municipal solid waste (MSW) gasification, the devolatilization submodel must be upgraded considering the heterogeneity nature of MSW residues, mainly composed of plastics (polyethylene, polystyrene, and polypropylene, among others) and cellulosic materials (cellulose, hemicellulose, and lignin). To solve this paradigm, our research group adopted a devolatilization submodel with the generation of secondary tar. Here, each of the MSW components is considered individually, following an Arrhenius kinetic rate. These MSW devolatilization equations and Arrhenius kinetic rates are detailed in Table 10. The full development of this MSW gasification model can be found in Ref. [23].

The ANSYS Fluent framework does not provide built-in devolatilization submodels for the Eulerian-Eulerian method, leading researchers to develop and implement in the Fluent code approaches such as the ones shown previously. Yet, built-in devolatilization submodel is available for the Eulerian-Lagrangian method. Four distinct devolatilization submodels are often employed. Regarding their implementation there are a few options to consider [7]:

• The constant rate model (usually set as the default model).

The constant rate devolatilization law dictates that volatiles are released at a constant rate A_0. The rate constant A_0 quantifies the rate of the chemical reaction and is defined in the material properties of the combusting particles. The Fluent materials database includes default values for each of the available combusting particle materials.

• The single kinetic rate model.

The devolatilization rate is first-order reliant on the amount of volatiles remaining in the particle. The kinetic rate, k, is defined by the input of an Arrhenius type preexponential factor and activation energy. For clarification purposes, the Arrhenius reaction rate relates the temperature dependence of a chemical reaction rate, and the activation energy is defined as the minimum quantity of energy that the reacting species must own to undergo a specified reaction.

Table 10 Biomass and MSW chemical reactions and devolatilization model equations [23, 73].

Reactions	Arrhenius reactions rate
Biomass devolatilization:	
$Biomass \rightarrow char + volatiles + steam + ash$	$r_1 = A_i \exp\left(\frac{-E_i}{T_s}\right)(1 - a_i)^n$
$Volatiles \rightarrow \alpha_1 CO + \alpha_2 CO_2 + \alpha_3 CH_4 + \alpha_4 H_2$	$r_2 = A_i \exp\left(\frac{-E_i}{T_s}\right)(1 - a_i)^n$
MSW devolatilization:	
$Cellulose \rightarrow \alpha_1 volatiles + \alpha_2 TAR + \alpha_3 char$	$r_3 = A_i \exp\left(\frac{-E_i}{T_s}\right)(1 - a_i)^n$
$Hemicellulose \rightarrow \alpha_4 volatiles + \alpha_5 TAR + \alpha_6 char$	$r_4 = A_i \exp\left(\frac{-E_i}{T_s}\right)(1 - a_i)^n$
$Lignin \rightarrow \alpha_7 volatiles + \alpha_8 TAR + \alpha_9 char$	$r_5 = A_i \exp\left(\frac{-E_i}{T_s}\right)(1 - a_i)^n$
$Plastics \rightarrow \alpha_{10} volatiles + \alpha_{11} TAR + \alpha_{12} char$	$r_6 = \left[\sum_{i=1}^{n} A_i \exp\left(\frac{-E_i}{RT}\right)\right]\rho_v$
$Primary\,TAR \rightarrow volatiles + Secondary\,TAR$	$r_7 = 9.55 \times 10^4 \exp\left(\frac{-1.12 \times 10^4}{T_g}\right)\rho_{TAR1}$
Homogeneous reactions:	
CO combustion: $CO + 0.5 O_2 \rightarrow CO_2$	$r_8 = 1.0 \times 10^{15} \exp\left(\frac{-16,000}{T}\right) C_{CO} C_{O_2}^{0.5}$
Water-gas shift: $CO + H_2O \rightarrow CO_2 + H_2$	$r_9 = 5.159 \times 10^{15} \exp\left(\frac{-3430}{T}\right) T^{-1.5} C_{O_2} C_{H_2}^{1.5}$
Reforming of hydrocarbons: $CO + 3H_2 \leftrightarrow CH_4 + H_2O$	$r_{10} = 3.552 \times 10^{14} \exp\left(\frac{-15,700}{T}\right) T^{-1} C_{O_2} C_{CH_4}$
H_2 combustion: $H_2 + 0.5 O_2 \rightarrow H_2O$	$r_{11} = 2780 \exp\left(\frac{-1510}{T}\right)\left[C_{CO} C_{H_2O} - \frac{C_{CO_2} C_{H_2}}{0.0265 \exp\left(\frac{3968}{T}\right)}\right]$
CH_4 combustion: $CH_4 + 2O_2 \rightarrow CO_2 + 2H_2O$	$r_{12} = 3.0 \times 10^5 \exp\left(\frac{-15,042}{T}\right) C_{H_2O} C_{CH_4}$
Heterogeneous reactions:	
Partial combustion: $C + 0.5 O_2 \rightarrow CO$	$r_{13} = 596\, T_p \exp - \left(\frac{1800}{T}\right)$
CO_2 gasification: $C + CO_2 \rightarrow 2CO$	$r_{14} = 2082.7 \exp - \left(\frac{18,036}{T}\right)$
Steam gasification: $C + H_2O \rightarrow CO + H_2$	$r_{15} = 63.3 \exp - \left(\frac{14,051}{T}\right)$

- The two competing rates model (The Kobayashi model).

ANSYS Fluent provides the kinetic devolatilization rate expressions, where competing rates control the devolatilization over different temperature ranges. The Kobayashi model requires the input of the kinetic rate parameters, A_1, E_1, A_2, and E_2, and the yields of the two competing reactions, α_1 and α_2.

• The chemical percolation devolatilization (CPD) model.

The CPD model characterizes the devolatilization behavior of rapidly heated coal based on the physical and chemical transformations of the coal structure. This model is extended to biomass devolatilization considering its major components' chemical structure and transformation under various mechanisms.

The literature is nonconsensual regarding which submodel approach is the most appropriate to deal with general gasification processes [2, 18]. Still, Silaen and Wang [70] comparatively studied these four devolatilization submodels applied to gasification processes. The main conclusions drawn showed that the Kobayashi model delivered the slowest devolatilization rates, while the constant rate model delivered the fastest. The single-rate model and the CPD model presented enhanced modeling behavior by producing moderate and consistent devolatilization rates, although the CPD model was more computationally expensive. A slower devolatilization rate prediction by the devolatilization submodel affects simulation results by providing lower H_2 production levels, higher outlet gas temperature, higher CO production and CO_2 mass fractions, and lower heating value and gasification efficiency. Thus, the devolatilization submodel can be rather impactful in several gasification conditions, therefore, one must consider the most appropriate devolatilization submodels to conserve the solution's accuracy and avoid misleading results.

5.4.2 Homogeneous gas-phase reactions submodels

The devolatilization step involves thermal cracking of molecular structures, while gasification involves the conversion to volatile products. During the gasification process, a set of chemical reactions take place, these can be divided into homogeneous and heterogeneous reactions. Chemical reactions processes occurring within the gasifier are rather complex to measure, given the difficulty in identifying all of the chemical species involved in the gasification reactions [83]. Approaches such as large numbers of detailed reactions involving radicals, or a set of global chemical reactions focusing on the main species generated, can be used to describe these reaction systems [37]. However, the former is excessively complex given the considerable number of processes that need to be modeled in a gasifier, therefore, the global homogeneous and heterogeneous reactions approach has been broadly used to model industrial gasifiers [67]. If desired, optimized chemical reactions kinetic mechanisms such as GRI-Mech 3.0 or USC Mech II may be additionally implemented in Fluent to achieve a detailed chemical kinetic modeling. These mechanisms operate by including a set of additional species and reactions involved in gasification processes [81]. However, the higher the number of chemical reactions considered, higher the computational demand will be. This requires the use of additional tools like In Situ Adaptive Tabulation (ISAT) which aids the incorporation of detailed chemical reactions mechanisms in multidimensional flow simulations by accelerating the chemistry calculations [38]. For the time being, a detailed chemical reactions kinetic model will not be considered in the scope of this implementation guide, instead, a simpler homogeneous and heterogeneous chemical reactions modeling approach will be presented.

The homogeneous reactions include only reactants and products in the same phase. In gasification, these reactions occur in the gas-phase describing the reactions between volatile gases and gasifying agents. To model these reactions, one should consider both the kinetic and the turbulent mixing rate effects. The gas-phase afflicted by the effects induced by the chaotic fluctuations of the solid particles lead to velocity and pressure fluctuations on the gaseous species. Typical homogeneous reactions considered are shown in Table 10. To deal with these reactions ANSYS Fluent provides a set of four reaction rate models dealing with the turbulence-chemistry interaction phenomena [7]:

- The laminar finite-rate model.

Determines reaction rates by Arrhenius kinetic expressions and ignores the effect of turbulent fluctuations. The model is accurate in describing laminar processes, but usually inaccurate for turbulent processes given the high nonlinear Arrhenius chemical kinetics.

- The eddy-dissipation model.

This is a turbulent-chemistry model in which the reaction rates are controlled by turbulence mixing, this way expensive Arrhenius chemical kinetic calculations can be avoided. The model is computationally cheap, however, for realistic results, only one or two-step heat-release mechanisms should be used.

- The combined finite-rate/eddy-dissipation model.

Considers both the Arrhenius and the eddy-dissipation reaction rates. The eddy-dissipation model assumes that reactions are fast and that the system is purely mixing-limited. Whenever that is not the case, it can be combined with finite-rate chemistry, thus the kinetic rate is calculated in addition to the reaction rate predicted by the eddy-dissipation model.

- The eddy-dissipation-concept (EDC) model.

The EDC model is an extension of the eddy-dissipation model and includes detailed chemical mechanisms in turbulent flows, yet, the detailed chemical kinetic calculations turn this model computationally expensive.

The eddy-dissipation model computes the turbulent mixing of the gases; however, it does not consider the limiting effect of temperature. On the other hand, the finite-rate kinetic model considers the reaction rate temperature but does not consider the turbulent mixing of the species. To amend this, the finite-rate/eddy-dissipation model combines the effect of turbulence on the reaction by computing both the Arrhenius and the eddy-dissipation mixing rates, providing the net reaction rate as the minimum of the two [7]. Therefore, the combined finite-rate/eddy-dissipation model is widely used to model the homogeneous gas-phase reactions in gasification processes [1, 17, 73].

5.4.3 Heterogeneous reactions submodels

In the heterogeneous reactions, the gas species react with solid char (the solid devolatilization residue) resulting in gaseous products. When compared with the other gasification process steps (devolatilization and homogeneous reactions), the heterogeneous reactions have a much slower reaction rate, thus they are considered as the controlling step during the whole gasification process.

Heterogeneous processes occur at the interfaces where different phases meet. In these interface sites, a set of chemical reactions take place, known as reactions on surfaces. These reactions involve more than one phase, in which at least one of the steps of the reaction mechanism is the absorption of one or more reactants. Table 10 provides the generally employed simplified heterogeneous reactions. To model these important interactions there are currently four heterogeneous surface reaction models provided within the ANSYS Fluent framework [7]:

- The diffusion-limited surface reaction rate model.

This is the default model in ANSYS Fluent and assumes that the surface reaction proceeds at a rate determined by the diffusion of the gaseous oxidant to the surface of the particle. Moreover, it assumes that the diameter of the particles remains unchanged. Once the mass of the particles is decreasing, the effective density decreases and the char particles become more porous.

- The kinetic/diffusion surface reaction rate model.

This model assumes that the surface reaction rate is determined either by Arrhenius rate kinetics or by a diffusion rate of the oxidant at the surface particle. Here, the particle size is assumed to remain constant while the density is allowed to decrease.

- The intrinsic model.

Similar to the kinetic/diffusion model, the intrinsic model assumes that the surface reaction rate includes the effects of both bulk diffusion and chemical reaction. However, the intrinsic model is more detailed and requires more fuel-specific input data information than the kinetic/diffusion model.

- The multiple surface reactions model.

This model follows a pattern similar to the wall surface reaction models, where the surface species is now a "particle surface species." Here, it is possible to set multiple particle surface reactions to model the surface combustion of a combusting discrete-phase particle.

As the heterogeneous reactions occur at the surface of solid particles the diffusion rate comes as an important process step to consider for certain reactions. The kinetic/diffusion surface reaction rate model is frequently used to simulate the heterogeneous surface reactions once it accounts for a diffusion-controlled regime and the solution kinetics for the boundary layer diffusion, allowing to evaluate the rate constants for the surface reactions [57, 73, 76].

5.5 Radiation model

Radiation is known as the transfer of heat through electromagnetic energy. In homogeneous media, such as in gas-rich environments, absorption and emission are the dominant mechanisms of radiation heat transfer, while in heterogeneous media scattering mechanisms are brought in due to the presence of inhomogeneities, such as flowing solid particles in fluidized bed reactors. Radiation heat transfer processes within gasifiers originate as emissions from the heat and energy transfer between different gas species, solid particles, and the gasifier's walls [34]. Thermal radiation effects should be accounted whenever the heat radiation is at least equal or of greater magnitude than that of convective and conductive heat transfer rates, being of practical importance only at very high temperatures (above 800 K) [82]. Radiation phenomena undergo complex interactions between the phases, so to accurately predict these interplay, computationally effective thermal radiation models are required to solve the radiative intensity transport equations. Table 11 summarizes the current five radiation models available to solve these equations.

Indeed, different models provide different solutions, since a given radiation model may be more appropriate to solve a certain problem than the other. Regarding

Table 11 Current applied radiation models and their applicability [7].

Model	Applicability
Discrete Ordinates (DO) Model	The DO model is the most comprehensive radiation model but can easily become extremely computationally expensive.
Discrete Transfer Radiation Model (DTRM)	The DTRM is a relatively simple model whose accuracy relies heavily on the number of rays; a large number of rays turns the problem-solving computationally intensive. It assumes all surfaces as diffuse without accounting for the scattering effects.
P-1 Model	In the P-1 model, the radiative equations can be easily solved with little computational cost but tend to overpredict radiative fluxes from localized heat sources. It includes the effect of scattering and can easily be applied to various complicated geometries.
Rosseland Model	The Rosseland model can only be applied to optically thick media. Its framework is derived from the P-1 model equations, with some approximations. It is faster than the P-1 model and requires less memory.
Surface-to-Surface (S2S) Model	The S2S model can be used in situations where there are no participating media (absorbs, emits, or scatters a thermal ray as it travels through the medium), however, it cannot be used with periodic or symmetry boundary conditions. Storage and memory requirements can be expensive with a large number of surface faces. It cannot be used with 2D axisymmetric geometries. When compared to DTRM and DO models, S2S has much faster computation time per iteration.

gasification modeling, based on literature review, the P-1 model has been widely applied to gasification simulation, mainly due to its robustness, stable operation, easiness, and little computational demand, while accounting for the radiation exchange between gas and particulates [51, 53, 60]. As for the remaining models, the Rosseland model may yield unreasonable and unrealistic results in predicting the gasification behavior, the S2S model does not consider participating media and DO and DTRM models are two times more computationally demanding than the remaining models [53, 60].

However, some researchers tend to apply a simpler approach and neglect the effect of thermal radiation on heat transfer in gasifiers. Thermal radiation effects may be negligible in gasification processes whenever heat losses by radiation are imperceptible [48, 56, 86]. Moreover, as only partial combustion occurs at the initial stage of gasification processes substantially high temperatures cannot be achieved, which withdraws practical importance in applying the radiation model [53]. Therefore, these assumptions neglect the radiation model effects on the final solution in gasification processes simulations. Lu et al. [53] delved into this issue and compared no radiation model application with the five different radiation models in gasification simulation. Fig. 13 provides the authors' syngas composition, the temperature at the outlet of the gasifier, and the average CPU (central processing unit) time per iteration for the predictions made without radiation model and with each one of the five radiation models. Results show that the predictions made for the syngas composition by all models are in fact very consistent. This agreement is also clear for the temperature predictions with the no radiation model application, P-1 model and DO model, while the Rosseland, DTRM, and S2S models are the greatest outliers.

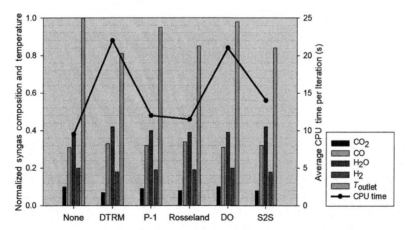

Fig. 13 Normalized syngas composition ($CO + H_2O + CO_2 + H_2 = 1$) and gasifier outlet temperature (T_{outlet}), and average CPU time per iteration (s) predictions without and with different radiation models applied to gasification simulation.

Additionally, simulation results also clarify the overdemanding computational cost for the DTRM and DO models. So, similar to previous sections, it all comes down to a compromise between accuracy and computational cost, however, this time the accuracy input by the radiation model to gasification processes may not always be worthwhile the trouble in implementing it. Thus, neglecting the radiation model allows researchers to simplify the analysis and save computational time, particularly if radiation phenomena are not mandatory to capture in the analysis at hands, namely in a syngas composition study.

6. Solution setup and calculation tasks

Having defined all the required setup settings and models, it is now time to solve all the corresponding governing equations for the conservation of mass, momentum, energy, and species (when appropriate) as well as all other scalars such as turbulence and chemical species.

Before proceeding with the solution calculation, the user must first set the solution methods. ANSYS Fluent provides multiple schemes to solve different types of solutions, however, only the available solution methods settings for solving Eulerian multiphase flows will be addressed. For Eulerian multiphase flows, ANSYS Fluent solves the phase momentum equations, the shared pressure, and phasic volume fraction equations either by implementing a coupled or a segregated fashion. When performing a pressure-based solver, one is entitled to choose the appropriate "Pressure-Velocity Coupling Method." Here, two distinct possible schemes are provided, "Phase Coupled SIMPLE" (suitable for Eulerian multiphase) and "Coupled" (suitable for all multiphase models). The Phase Coupled SIMPLE algorithm has proven to be robust and successful in determining a broad range of multiphase flows, solving the velocity equations coupled by phases in a segregated manner [7]. On the other hand, the Coupled algorithm employs a coupled mode, solving all equations for phase velocity corrections and shared pressure correction simultaneously, showing improved efficiency in dealing with steady-state situations or transient simulations when larger time steps are needed. Both schemes are suitable for performing fluidized bed gasification simulations.

Regarding the Spatial Discretization scheme, in Fluent, one can choose how cell-centered gradients are calculated. Gradients are needed to deduce the values of a scalar at the cell face and to compute the secondary diffusion terms and velocity derivatives. ANSYS Fluent provides three possible gradients, Green-Gauss Cell-Based, Green-Gauss Node-Based, and Least-Squares Cell-Based (default method). In short, Least-Squares and Green-Gauss Node-Based are the most accurate, however, Least-Squares is slightly cheaper than Node-Based method. On the other hand, Green-Gauss Cell-Based is least accurate but much cheaper than the previous two. Here, the Least-Squares Cell-Based method is widely applied since it is appropriate for any kind of mesh.

For each scalar equation listed under "Spatial Discretization" namely, momentum, volume fraction, energy, turbulent kinetic energy, turbulent dissipation rate, among others possible appearing features (depending on the model options previously set), the user must choose an interpolation method, those being: first-order upwind, second-order upwind, power law, QUICK (quadratic upwind interpolation), and third-order MUSCL (monotone upstream-centered schemes for conservation laws). Without getting into much detail regarding each interpolation method, first-order upwind is the default discretization method for multiphase flows, being the easiest to converge but only first-order accurate, while the second-order upwind provides more accurate results, albeit convergence may be slower. Considering these assumptions, one may choose which method is more stable for the problem in hands, in the event of a certain scalar initially defined as second-order might behave unstably during the calculation, the user may interrupt the process and set that specific scalar as first-order and resume with the simulation. Generally, choosing between these two discretization methods provide the right amount of accuracy needed to solve most problems (gasification processes included). Additionally, both first- and second-order upwind are the most recommended methods for beginners. As for the remaining methods, Power Law has some niche utility, being only slightly better than first-order and not as accurate as second-order when all types of flow scenarios are considered; QUICK and third-order MUSCL are the most accurate (both third-order), but also the most intricate to apply and converge, requiring high-quality grid and improved setting adjustments, are therefore recommended for more advanced users.

After setting the discretization scheme, the user must define the numerical algorithm for the transient term, here known as "Transient Formulation." ANSYS Fluent disposes of three implicit time-stepping schemes, first-order implicit, second-order implicit, and bounded second-order implicit (available only for the pressure-based solver). The first-order implicit, already set by default, is the most stable of the three, being usually sufficient for most problems, however, it is the less accurate. For higher accuracy requirements, the user must consider either the second-order implicit (suitable for stable 2D gasification processes) or the bounded second-order implicit. Both second-order schemes provide the same accuracy, yet, the bounded second-order implicit grants better stability, since time discretization ensures the bounds for variables, being recommended for more complex solutions with difficulty in converging (such as complex 3D gasification processes).

Within the solution controls tab the user can adjust the underrelaxation factors. These factors are used to turn the solution process more stable by preventing large changes between solution values from one iteration to the next so that they do not affect the accuracy of the final solution, only it takes to reach the number of iterations. The underrelaxation factors are particularly important during the initial stages of the simulation, in which large fluctuations of the solution variables may occur. After this initial stage, the solution flow settles down and heads toward stability, from this moment on, under-relaxation factors should not be as mandatory. It is good practice to initiate the calculation using the default underrelaxation factors, however, in order to achieve convergence, users sometimes have to go through a trial and error effort. Setting lower underrelaxation factors helps the convergence process, however, it may dramatically delay the simulation time, since setting very low factors will make the variables to change by a very small amount. Additionally, if the user has to set very low underrelaxation factors in order to achieve convergence, then something might be wrong with the solution, thus a general review of all the previous settings made is avidly recommended. Therefore, regarding underrelaxation factors settings, optimal values can be somewhat problem dependent and are better learned from experience.

During the simulation calculation, one can monitor the convergence process by plotting, for each solver iteration, data concerning quantities of interest during the solution. Properties such as residual values, statistics, force values (drag, lift, and moment coefficients), surface integrals (set at a boundary or at any other defined surface of the domain), and volume integrals (set at cell zones), can be computed and stored, recording the convergence history of the solution.

At this point, the solution is ready for initialization. Properly setting the initialization values is a rather important part of the process, as realistic initial guesses improve the solution stability and accelerate convergence, otherwise poor initial guesses may compromise the solution failing within the first few iterations. ANSYS Fluent provides two main initialization methods, Hybrid and Standard Initialization. The former makes nonuniform initial value guesses for every individual cell in the mesh, which is often useful to simulate in a more general environment or normal conditions; while the latter allows the user to specify each and every variable value (gauge pressure, air and solids temperature, air and solids velocity, and others), initializing the mesh cells with the values assigned by the user. Regarding gasification processes, as these are commonly set to work within very specific controlled environments then the user should employ a standard initialization for better specification of the problem environment. As most of these variable parameters have been already inputted during the boundary conditions settings, users may use the "Compute from" option, enabling to automatically fill in the initialization values with the previously specified values.

After the initialization the "Patch" button becomes enabled. Setting the Patch values for individual variables in certain regions of the domain is an essential task while modeling multiphase flows and combustion problems. In order to do so, one must first create a domain region adaption, within the Adapt panel, which marks individual cells for refinement. In fluidized bed simulation, this marked cell region of interest is the area occupied by the static bed (delimited by the extremities of the reactor's domain and the bed height), this selection is very important to set the different phase volume fractions in the region. Having defined the region for adaption, it is a good practice to display it so to visually verify if it encompasses the desired area. Following this procedure, the user may now return to the Patch panel and set the initial volume fraction of the solids (bed material and biomass in case of a binary mixture) in the marked bed region of the fluidized bed.

Before proceeding with the solution calculation, the user must set the "Calculation Activities," herein one can establish the number of iterations at which ANSYS Fluent will autosave the simulation results, jointly with other tasks such as exporting files (here the user may select the quantities to be exported to the postprocessing), creating solution animations, and command execution (defined to generate animation frames such contour plot and vector plots).

The "Run Calculation" allows to finally start the solver iterations. The available panel options in this task page vary concerning previous settings made. For a transient flow calculation, the user disposes of various options to determine the time step. Employing a "Fixed" time-stepping method allows the user to input the intended time step size (in seconds) and the number of time steps. If desired, the user may enable the "Extrapolate Variables" option, to estimate an initial guess for the next time step, allowing to speedup the transient solution by reducing required subiteration; and the "Data Sampling for Time Statistics," to compute the time-average (mean) and root-mean-square of the instantaneous values sampled during the calculation. All remaining options may be set as default.

Additionally, during the calculation process, the user may display contours, vectors, monitor plots or even mesh, for any desired quantity in the "Graphics and Animations" drop-down list. Displaying the solids volume fraction contours in the reactor's domain is particularly useful for hydrodynamic analysis in fluidized bed gasification, allowing the user to follow the solids evolution being refreshed for every time step throughout the simulation time. Having finished all these solver settings, the user may now start the calculation.

In order to recap all the main settings discussed throughout this section, an outline of the general procedure on how to properly set the solver in ANSYS Fluent is presented:

1. select the pressure-velocity coupling method;
2. choose the discretization schemes and transient formulation;
3. set the underrelaxation factors;
4. enable the appropriate solution monitors;
5. initialize the solution;
6. adapt region and patch values;
7. display contours for the desired quantity (i.e., solids volume fraction);
8. start calculation.

As seen, each of these steps is composed of a large number of parameters that govern solver performance and accuracy. Although the most important settings regarding gasification processes were broadly covered throughout this implementation guide, there are some less impactful remaining settings that were not discriminated, for those, usually, the

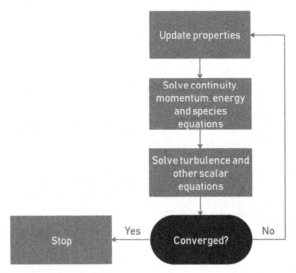

Fig. 14 Roadmap to convergence.

default values are good enough to ensure a robust path to a converged simulation. Fig. 14 gives a simplified vision of that exact path.

As shown in the above framework, the discretized conservation equations are solved until convergence is reached using an iterative process. This process ends when changes in solution variables from one iteration to the next are negligible. The overall accuracy of a given converged solution will mainly depend on all the previously studied parameters, for instance, the appropriateness and accuracy of physical models, assumptions made and the mesh resolution, and on the numerical errors made.

Solving a multiphase flow system can be inherently complicated and users may face some stability or convergence problems. Thus, occasionally it is necessary to modify the solver parameters (i.e., numerical schemes, relaxation factors, time step, among others) in order to achieve better convergence, stability, and higher accuracy. Therefore, it is so important to correctly set all the preceding stages of the modeling process discussed throughout this implementation guide, to avoid further instabilities arising from poor initial settings. Instead of going through all of these parameters and explaining each one individually on how to set them and when to use them, a few tips on how to properly set the solvers for gasification cases are provided:

- If the user desires to shorten the computational time for transient solutions, the Phase Coupled SIMPLE scheme is more appropriate than the coupled scheme.
- In the event of a divergent behavior is persistently achieved while applying a Phase Coupled SIMPLE scheme, replacing it for a Coupled scheme may sometimes aid the convergence, granting a steadier solution which is useful particularly when dealing with more intricate geometries.

- In case the user chooses to enforce a couple scheme, a courant number (200 by default), and explicit relaxation factors for momentum and pressure (0.75 by default) must be specified. The user may decrease the courant number and the explicit relaxation factors if difficulties in reaching the convergence are encountered, whether being due to higher-order schemes or to the high complexity of the problem, such as in multiphase and combustion problems. These can later be heightened if the iteration process runs smoothly.
- Different numerical schemes may respond differently to the applied underrelaxation factors. For instance, setting lower underrelaxation factors for the volume fraction equation for the coupled scheme may lag the solution considerably, values placed around 0.5 or above are considered acceptable. Contrarily, the Phase Coupled SIMPLE generally requires a low underrelaxation for the volume fraction equation.
- When the solution solver requires higher-order numerical schemes or higher spatial discretization, it is recommended that the user initiates the solution by setting smaller time steps. These may be further increased after performing a few time steps so to achieve a better approximation of the pressure field.
- For better convergence in gasification analysis, it is recommended to start the solution with a nonreacting flow and without radiation model. To do so, it is necessary to disable the chemical reactions, radiation equations, and fluid-particle interactions. For instance, if the user intends to evaluate the solid particles behavior within a fluidized bed, the solution must be initiated without the inclusion of the chemical reactions submodel. This allows analyzing, in a first stage, the hydrodynamics features by employing a simpler approach, determining if the results obtained are within tolerance and if proper behavior is being achieved. After such validation, the chemical reactions model and the devolatilization submodel can be securely added to the mathematical model.
- To "kick-start" the simulation, it is recommended to patch a higher temperature in the reaction region or patch intermedium species which could also help with this regard.
- When applying the DPM, while performing a single iteration with continuous phase iteration per DPM iteration, setting the underrelaxation factor to 1 before bringing it to the desired value can help achieve better convergence.
- When running a reacting flow, set lower species and energy equation with underrelaxation factors placed around 0.9 and gradually ramp them up to 1.
- For nonconverging DPM cases, reduce the DPM underrelaxation factor to 0.1 or lower.
- Finally, the residuals stand as useful indicators of the iterative converge of the solution, quantifying the error in the solution of the equations system, thus it is important to motorize the residuals behavior during the calculation. Throughout this iterative process, the residuals are expected to progressively decay to smaller values (never reaching exact zero) up until they get leveled and substantial changes stop occurring. The lower the residual values are, the more numerically accurate the solution will be.

Table 12 Tips and tricks for solution setting and calculation tasks.

User problem	Solution
Slow convergence.	Setting better initial conditions; progressively raising underrelaxation factors or Courant number; and initializing the solution with a good quality mesh with appropriate mesh resolution, all these tasks will help speeding-up the convergence process.
Struggling with the results accuracy.	The accuracy relies mostly upon implementing high-order discretization schemes, mesh resolution, boundary conditions, model limitations (chemical reactions, devolatilization, turbulence, radiation, etc.), geometry simplifications, among other features.
Bad quality mesh.	One can improve the final numerical mesh quality in ANSYS Fluent by smoothing and face swapping the mesh, and also by setting a polyhedral mesh conversion. These tasks can be performed in the "Mesh" task page.

If the residuals present an increasing behavior within the first few iterations, one should consider lowering the underrelaxation factors and resume the calculation. Occasionally, the residuals may present a rather unstable behavior showing huge fluctuations, on such occasions the user should proceed by reducing the underrelaxation factors. Yet, if the instabilities prevail this might be a sign of previous misconfigurations, thereby, the user must recheck previous settings such as initial values, boundary conditions, mesh and fluid properties, in order to reach more stable residual curves.

Table 12 provides some additional tips and tricks to help sidestep some common obstacles while setting up the solver.

7. Model validation

Having gone through all this tiresome process to obtain a final converged solution there is a final step that researchers must endure, revising their results. Getting a converged solution does not mean that said solution faithfully represents the physical process being studied. There are two systematic key processes for confirming numerical results, verification, and validation. The verification process consists in determining if the implemented computational model accurately represents the user's conceptual description of the model, in other words, it concerns if the equations were properly solved attending to the consistency of the mathematical model and the level of numerical errors. On the other hand, the validation process determines how far the implemented model is an accurate representation of the real world, carrying for the comparison of the numerical results with experimental data, or with reliable data gathered from the literature (useful in the event of the user does not possess experimental data for comparison) [78]. Thus, validation is an essential part of the process when analyzing the solution, as it is the only proper way to ensure the accuracy of the results obtained with the mathematical model.

In gasification, these are crucial processes to consider for the successful outcome of the solution since one is dealing with an extremely complex multiphase model. To proceed with the verification, the user must inspect if the solution is cohere based on engineering judgment, that is, if the flow features are reasonable and follow the expected trends of the theoretical model, or if one should reconsider the chosen boundary conditions and physical models:

- Perhaps there are 3D effects that are not being accounted for a 2D study?
- Considering a laminar case when in fact the flow is of turbulent nature?
- Opting by model fluids as incompressible in cases where there are compressibility effects?
- Selecting a steady state in an unsteady problem?
- Is the created domain large enough so it does not affect the results?
- Are the values chosen as boundary conditions realistic enough?
- Is this a grid independent solution?

Having settled a verification strategy and tested the uncertainty of the solution one must now advance with the validation process. In truth, the solution verification should precede its validation as the outcome of the validation process is pointless without identifying and quantifying the effects of numerical errors and their propagation on the model. For the model validation procedure, one must focus on the numerical results and assess if these are consistent with the experimental data, or with trustworthy literature. To attain a strong validation and guarantee the robustness and predictability of the mathematical model, it is of paramount importance to perform validation at various points and experimental conditions before interpreting the numerical results with confidence. Such methodology has been thoroughly employed by the research group [13, 30, 73].

Fig. 15 summarizes this validation process applied to the two pilot-scale fluidized bed reactors, 250 and 75 kW_{th}. The relative deviation between the experimental and the numerical syngas composition for a set of gasification runs performed with forest residues and coffee husks in the 250 kW_{th} reactor is depicted in Fig. 15A. Details on the experimental operating conditions can be found elsewhere [13]. Results show that the numerical model successfully predicted the trends for the syngas compositions for the six experimental runs with a close agreement. The largest deviations were measured for the smaller gaseous fractions, methane (CH_4) and hydrogen (H_2), with around 17% and 16%, such is due to their smaller volume fraction in the gas favoring higher relative errors.

Fig. 15B exhibits the deviation between the experimental and the numerical fluidization curves retrieved at two different bed heights (8 and 18 cm) from the 75 kW_{th} reactor. Operating conditions were set at a temperature of 873 K for various superficial gas velocities and simulation time of 3 s. Overall, the numerical curves successfully predicted

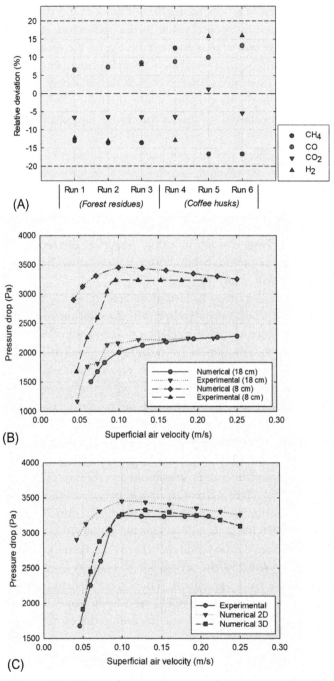

Fig. 15 Model validation: (A) Relative deviation between the experimental and numerical syngas composition for a set of six gasification runs performed with forest residues (three runs) and coffee husks (three runs) in the 250 kW$_{th}$ reactor; (B) experimental and numerical fluidization curves gathered at 8 and 18 cm height from the 75 kW$_{th}$ reactor; (C) experimental, 2D and 3D numerical fluidization curves for the 75 kW$_{th}$ reactor.

the experimental curves slope with good agreement. The larger deviations measured for the lowest velocities are given to the negligible solids movement before fluidization occurs, and to the inefficiency of the mathematical model within the ANSYS Fluent database to consider such low entropy behavior [11]. The higher-pressure range measured for the 8 cm height curve is due to the close proximity to the turbulent air inlet as compared to the 18 cm curve.

To extend the validation process to a 3D simulation, Fig. 15C presents the deviation between the experimental and the 2D and 3D numerical fluidization curves for the 75 kW$_{th}$ fluidized bed reactor. Again, an operating temperature of 873 K for various superficial gas velocities and a simulation time of 3 s were set for both geometries. Results show that the 2D simulation tends to overestimate the fluidization curve behavior, while the 3D simulation approaches more closely to the experimental fluidization curve. The 3D simulation provides lower pressure values as the reactor wall area and volume ratio is larger for the 3D configuration, justifying the increased pressure drop prediction presented by the 2D simulation [21]. Once again, the 3D simulation shows an enhanced capability to deliver a more accurate and realistic behavior analysis than the 2D simulation.

All numerical results here presented already comprise the mathematical model enhanced with UDF routines. The implementation of UDF into the ANSYS Fluent framework is rather advantageous validation-wise. As previously mentioned, these allow tuning the solution to the user's modeling needs, in this particular case to the reactors geometry, boundary conditions and material properties, allowing to strengthen the validation process by minimizing the deviations.

Indeed, the mathematical model was capable to effectively predict the acquired experimental data trends with generally good agreement for both gasifying units, at various validation points and experimental conditions. The model applied here has already been extensively validated and submitted to constant improvements in dealing with different biomass substrates [14, 31] and the heterogeneity of MSW [11, 23] at different operating conditions [15, 24], gasifying agents [25, 27], and reactor scales [13, 26] since its first application and development made by Silva et al. [73].

Having gone through this process successfully by applying an effective verification and validation strategy, the user may now acknowledge the polyvalence potential of the mathematical model at hands in dealing with various scenarios and recognize the sufficient robustness and predictability required to achieve confident modeling results.

8. Postprocessing

The solution once converged and validated, the next logical step is to process the said solution to obtain the results first planned in the preanalysis step. Researchers enjoy a wide range of options when it comes to postprocessing such as contour plots, vectors, streamlines, isosurfaces, video screenings, create planes and lines to study particular solution regions, graphical representations, and generate reports. Researchers dispose of various processing software like ANSYS CFD-Post, Tecplot 360, Ensight (now part of ANSYS universe), FieldView, ParaView, just to name a few. Within the ANSYS framework, there are two possible routes to postprocess the simulation results from Fluent, the Fluent postprocessing integrated tools, or the ANSYS CFD-Post application.

Some of the aforementioned postprocessing options can be directly applied in the ANSYS Fluent postprocessing. The tools built into Fluent have the advantage of letting the user to promptly review the simulation data, or even stop the simulation to examine the results, and further modify and resume with the calculations if required. Postprocessing tasks performed within the Fluent solver are often more convenient for most basic postprocessing operations. Within the "Results" task page, a set of most common postprocessing features can be found namely contours, vectors, pathlines, particle tracks, animations, several types of plotting and reports. Supplementary postprocessing tasks can be found within the "Postprocessing" banner in the toolbars tabs. Here, the user may access and create surface regions in the solution domain such as points, lines, and planes (the 3D time-averaged solids volume fraction contours shown in Fig. 7 stand as an example of this plane option use). In fluidized bed gasification, these sorts of features are rather useful to study a certain region within the reactor's vessel, or to study a desired variable behavior, for instance, one may plot the biomass velocity in the near-wall region, or the bed expansion by plotting the solids volume fraction by means of a centerline along the reactor's height, or even probe the static pressure in any location of the reactor the user sees fit.

The other possible route for postprocessing is to use the ANSYS CFD-Post. Fig. 16 shows an overview of the ANSYS CFD-Post user interface. Both platforms are perfectly

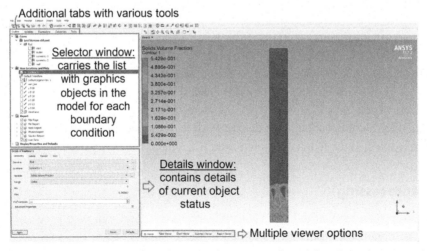

Fig. 16 ANSYS CFD-Post user interface depicting the contours for solids volume fraction within the fluidized bed reactor.

capable of addressing the most basic postprocessing, however, contrarily to the ANSYS Fluent postprocessing built-in options, the CFD-Post provides far more powerful and sophisticated postprocessing capabilities as 3D-viewer files, user variables, automatic HTML report generation, and case comparison tools [8]. ANSYS Fluent allows the user to send the case and data files to CFD-Post, to perform the various postprocessing actions. For such the user must select the quantities one desires to export by creating a "CFD-Post Compatible Automatic Export" within the "Calculation Activities" task page. Regarding fluidized bed gasification, the use of CFD-Post is advantageous as it allows to produce visual data with higher quality, assisting to better visualize and understand the complex flow phenomena within the reactor. Indeed, the ANSYS CFD-Post options are immense and are best learned in a hands-on manner, the experience will lead researchers to take their results visualization and analysis to the next level by taking the maximum potential of such a broad application.

9. Conclusions, limitations, and future prospects

This tutorial guide has summarized and discussed the main process steps one must endure to implement a gasification model within the ANSYS Fluent framework. Fig. 17 provides the last rundown of the process steps discussed throughout this tutorial that one

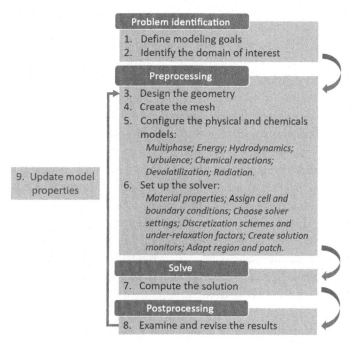

Fig. 17 Complete solution procedure overview.

must engage to fully perform a gasification analysis. From the problem identification up to the postprocessing, the main phenomena occurring within the reactor are covered and supported by practical application examples from recent work developed by the research group within the field. Some of the main current challenges regarding gasification CFD analysis such as the scaling-up, 2D vs 3D solutions, solver customization by implementing UDF for better results, and numerical model validation with experimental data gathered from two different pilot-scale fluidized bed reactors (75 and 250 kW_{th}) are also addressed. Along the way, some tips and tricks are delivered, aiding researchers to better configure their solution. Overall, the CFD solver application developed showed to be sufficiently robust to deal with the multiple features to which it was submitted, proving to be an essential tool for the design and to understand the behavior of gasifying units submitted to multiple operating conditions.

However, just like any other computational technique enforced to solve real-life issues, CFD carries a set of limitations that users must be aware of and understanding the extent of these limitations is crucial for the model applicability in scientific and engineering problems. CFD solutions rely upon physical models developed to describe real-world processes namely turbulence, chemistry, multiphase flow, compressibility, among other phenomena. The solutions provided through CFD analysis are only as accurate as the physical models on which they are based. When applying CFD analysis to describe such a physically and chemically complex process as gasification, certain simplifications need to be incorporated and cannot be ignored during the modeling process. These all introduce limitations that must be acknowledged while performing a CFD gasification analysis:

- As gasification analysis becomes even more detailed and complicated, it is of sheer importance to control the numerical solutions through experimentation processes that allow bringing the solution closer to a real scenario such as in industrial-scale reactors. However, there are not many experimental data on reactors of this size, which makes it sometimes difficult to appropriately validate numerical solutions at this range.
- An alternative to circumvent this issue is to perform experiments on smaller-scale reactors, yet, a set of scale-dependent parameters change drastically when moving to smaller units such as hydrodynamics, heat transfer, residence time, which in turn will affect chemical reactions leading to noticeable syngas compositions changes [65]. Thus, the information resulting from smaller scales need to be extrapolated to a full scale and general rules for doing so rely, once more, upon engineering approximations and modeling shortcuts, increasing the error and uncertainty of the analysis.
- Experimentally wise, the manner by which biomass (or solid waste) is fed into the reactor will lead to variations in the relevant parameters profile within the system since it will cause biomass/solid waste to run differently with the medium [46]. The feed-in process variations and how these affect the gasification process are phenomena that are not thoroughly explored in CFD analysis and the importance in understanding this process step by means of numerical solutions is certainly a limitation in this regard.

- The Eulerian-Lagrangian approach, when applied to gasification processes, allows delivering a more precise resolution of the problem as it offers a more accurate description regarding particle motion, interactions, and chemical reactions. Nevertheless, limitations arise from this method as a high number of individual particles may be required for an accurate prediction of the flow field, being at the same time more computationally demanding. On the other hand, setting a lower number of individual particles reduces the accuracy considerably, but turns the solution more computationally feasible. However, it would be computationally prohibitive to employ a Euler-Lagrange approach to simulate the chemically reacting gas–solid flow within an industrial size reactor. Therefore, researchers are still constantly obliged to find viable tools to simulate larger systems with realistic operating conditions within a desirable accuracy.
- Up to this day, computational capabilities and numerical models are still unfit to provide a full resolution to the turbulent dispersed flow phase for real-world applications. Describing the particles motions and collisions within a fluidized bed gasifier is still a challenge given the shape and size-dependent motion of the particles and their complex interactions with the turbulent gas-phase carrier. Thus, to model these phenomena certain simplification assumptions are commonly applied such as assuming average particle diameter distributions and strictly spherical shaped particles.
- As seen throughout this implementation guide, occasionally, ANSYS Fluent built-in model options are limited, not providing the required amount of accuracy needed to fulfill the user's particular modeling needs, being necessary to implement UDF to customize the solution and bring it closer to a more realist behavior scenario.
- Lastly, most of the ANSYS Fluent built-in models applied to gasification were initially developed to deal with combustion processes, however, as gasification and combustion mechanisms strongly relate with each other these have been widely used to describe gasification processes. Nevertheless, users should be aware of the resemblances and differences between both processes, once depending on the type of analysis one wishes to perform it might be necessary to import into ANSYS Fluent required gasification mechanism steps to bring the solution closer to a gasification procedure.

Computers are assuming a progressively more important role in helping us describe and better understand the most complex real-world physical mechanisms. To this end, CFD stands as an extremely powerful tool in predicting the highly complex fluid flow physics. Moreover, CFD capabilities evolve hand-in-hand with the ever-growing computational power, allowing researchers to achieve even more effective and robust modeling solutions to describe the most intricate phenomena in nature. To this end, CFD tools may provide a helping hand to the scientific and engineering community to the study and optimization of sustainable solutions for the development of renewable energies solutions. Regarding gasification processes, CFD future trends are expected to work around its very own limitations, with constant improvements in code verification and experimental validation, alongside with the further development of chemical and physical models. In due time, this will allow CFD analysis to become an increasingly more reliable, efficient, and ultimately a fundamental tool in the design, analysis, and optimization of multiphase flow systems for numerous engineering applications.

References

[1] N.A. Abaimov, A.F. Ryzhkov, Development of a model of entrained flow coal gasification and study of aerodynamic mechanisms of action on gasifier operation, Therm. Eng. 62 (11) (2015) 767–772.

[2] I. Adeyemi, I. Janajreh, T. Arink, C. Ghenai, Gasification behavior of coal and woody biomass: validation and parametrical study, Appl. Energy 185 (2017) 1007–1018.

[3] A.A. Ahmad, N.A. Zawawi, F.H. Kasim, A. Inayat, A. Khasri, Assessing the gasification performance of biomass: a review on biomass gasification process conditions, optimization and economic evaluation, Renew. Sust. Energ. Rev. 53 (2016) 1333–1347.

[4] T.Y. Ahmed, M.M. Ahmad, S. Yusup, A. Inayat, Z. Khan, Mathematical and computational approaches for design of biomass gasification for hydrogen production: a review, Renew. Sust. Energ. Rev. 16 (4) (2012) 2304–2315.

[5] Z.A.B.Z. Alauddin, P. Lahijani, M. Mohammadi, A.R. Mohamed, Gasification of lignocellulosic biomass in fluidized beds for renewable energy development: a review, Renew. Sust. Energ. Rev. 14 (9) (2010) 2852–2862.

[6] ANSYS, Modeling Turbulent Flows, (2006).

[7] ANSYS, ANSYS Fluent Theory Guide, Release 15.0, ANSYS, Inc., 2013.

[8] ANSYS, ANSYS CFD-Post User's Guide, Release 15.0, ANSYS, Inc., 2013.

[9] D. Baruah, D.C. Baruah, Modeling of biomass gasification: a review, Renew. Sust. Energ. Rev. 39 (2014) 806–815.

[10] G. Bhutania, P.R. Brito-Parada, J.J. Cilliers, Polydispersed flow modelling using population balances in an adaptive mesh finite element framework, Comput. Chem. Eng. 87 (2016) 208–225.

[11] J. Cardoso, V. Silva, D. Eusébio, P. Brito, Hydrodynamics modelling of municipal solid waste residues in a pilot scale fluidized bed reactor, Energies 10 (2017) 1773.

[12] J. Cardoso, V. Silva, D. Eusébio, P. Brito, Hydrodynamic modelling of municipal solid waste residues in a pilot scale fluidized bed reactor, Energies 10 (11) (2017) 17773.

[13] J. Cardoso, V.B. Silva, D. Eusébio, L. Tarelho, P. Brito, M. Hall, Comparative scaling analysis of two different sized pilot-scale fluidized bed reactors operating with biomass substrates, Energy 151 (2018) 520–535.

[14] J. Cardoso, V. Silva, D. Eusébio, P. Brito, L. Tarelho, Improved numerical approaches to predict hydrodynamics in a pilot-scale bubbling fluidized bed biomass reactor: a numerical study with experimental validation, Energy Convers. Manag. 156 (2018) 53–67.

[15] J. Cardoso, V. Silva, D. Eusébio, P. Brito, R.M. Boloy, L. Tarelho, J.L. Silveira, Comparative 2D and 3D analysis on the hydrodynamics behaviour during biomass gasification in a pilot-scale fluidized bed reactor, Renew. Energy 131 (2019) 713–729.

[16] J. Chang, J. Zhao, K. Zhang, J. Gao, Hydrodynamic modeling of an industrial turbulent fluidized bed reactor with FCC particles, Powder Technol. 304 (2016) 134–142.

[17] S. Chapela, J. Porteiro, M. Costa, Effect of the turbulence-chemistry interaction in packed-bed biomass combustion, Energy Fuel 31 (2017) 9967–9982.

[18] W.-H. Chen, C.-J. Chen, C.-I. Hung, C.-H. Shen, H.-W. Hsu, A comparison of gasification phenomena among raw biomass, torrefied biomass and coal in an entrained-flow reactor, Appl. Energy 112 (2013) 421–430.

[19] P.N. Ciesielski, G.M. Wiggins, J.E. Jakes, C.S. Daw, Simulating biomass fast pyrolysis at the single particle scale, in: R.C.B.K. Wang (Ed.), Fast Pyrolysis of Biomass: Advances in Science and Tehcnology, The Royal Society of Chemistry, 2017.

[20] S. Cloete, S.T. Johansen, S. Amini, Performance evaluation of a complete Lagrangian KTGF approach for dilute granular flow modelling, Powder Technol. 226 (2012) 43–52.

[21] S. Cloete, S.T. Johansen, S. Amini, Investigation into the effect of simulating a 3D cylindrical fluidized bed reactor on a 2D plane, Powder Technol. 239 (2013) 21–35.

[22] N. Couto, V. Silva, E. Monteiro, P. Brito, A. Rouboa, Using an Eulerian-granular 2-D multiphase CFD model to simulate oxygen air enriched gasification of agroindustrial residues, Renew. Energy 77 (2015) 174–181.

[23] N. Couto, V. Silva, E. Monteiro, S. Teixeira, R. Chacartegui, K. Bouziane, P.S. D. Brito, A. Rouboa, Numerical and experimental analysis of municipal solid wastes gasification process, Appl. Therm. Eng. 78 (2015) 185–195.

[24] N. Couto, E. Monteiro, V. Silva, A. Rouboa, Hydrogen-rich gas from gasification of Portuguese municipal solid wastes, Int. J. Hydrog. Energy 41 (25) (2016) 10619–10630.

[25] N.D. Couto, V.B. Silva, A. Rouboa, Assessment on steam gasification of municipal solid waste against biomass substrates, Energy Convers. Manag. 124 (2016) 92–103.

[26] N. Couto, V.B. Silva, C. Bispo, A. Rouboa, From laboratorial to pilot fluidized bed reactors: analysis of the scale-up phenomenon, Energy Convers. Manag. 119 (2016) 177–186.

[27] N. Couto, V. Silva, A. Rouboa, Municipal solid waste gasification in semi-industrial conditions using air-CO2 mixtures, Energy 104 (2016) 42–52.

[28] N.D. Couto, V.B. Silva, A. Rouboa, Thermodynamic evaluation of Portuguese municipal solid waste gasification, J. Clean. Prod. 139 (2016) 622–635.

[29] N. Couto, V. Silva, J. Cardoso, A. Rouboa, 2nd law analysis of Portuguese municipal solid waste gasification using CO_2/air mixtures, J. CO_2 Util. 20 (2017) 347–356.

[30] N. Couto, V. Silva, E. Monteiro, A. Rouboa, Exergy analysis of Portuguese municipal solid waste treatment via steam gasification, Energy Convers. Manag. 134 (2017) 235–246.

[31] N. Couto, V. Silva, E. Monteiro, A. Rouboa, P. Brito, An experimental and numerical study on the Miscanthus gasification by using a pilot scale gasifier, Renew. Energy 109 (2017) 248–261.

[32] C. Dinh, C. Liao, S. Hsiau, Numerical study of hydrodynamics with surface heat transfer in a bubbling fluidized-bed reactor applied to fast pyrolysis of rice husk, Adv. Powder Technol. 28 (2017) 419–429.

[33] A.M. Eaton, L.D. Smoot, S.C. Hill, C.N. Eatough, Components, formulations, solutions, evaluation, and application of comprehensive combustion models, Prog. Energy Combust. Sci. 25 (4) (1999) 387–436.

[34] M. Eriksson, M.R. Golriz, Radiation heat transfer in circulating fluidized bed combustors, Int. J. Therm. Sci. 44 (2005) 399–409.

[35] E. Esmaili, N. Mahinpey, 3D Eulerian Simulation of a Gas-Solid Bubbling Fluidized Bed: Assessment of Drag Coefficient Correlations, WIT Press, 2009.

[36] H. Fan, D. Guo, J. Dong, X. Cui, M. Zhang, Z. Zhang, Discrete element method simulation of the mixing process of particles with and without cohesive interparticle forces in a fluidized bed, Powder Technol. 327 (2018) 223–231.

[37] R. Fernando, Developments in Modelling and Simulation of Coal Gasification, International Energy Agency (IEA) Clean Coal Center, 2014.

[38] S.B. Fiveland, C. Rutland, The development and use of rate-constrained chemical equilibrium with HCCI combustion modelling, Prepr. Pap. - Am. Chem. Soc., Div. Fuel Chem. 49 (1) (2004) 269–271.

[39] H.S. Fogler, N.M. Gurmen, Aspen Plus Workshop for Reaction Engineering and Design, The University of Michigan, 2002.

[40] V. Garzó, J.W. Dufty, C.M. Hrenya, Enskog theory for polydisperse granular mixtures. I. Navier-Stokes order transport, Phys. Rev. E 76 (2007) 031303.

[41] W. Godlieb, N.G. Deen, J.A.M. Kuipers, A discrete particle simulation study of solids mixing in a pressurized fluidized bed, in: 2007 ECI Conference on the 12th International Conference on Fluidization—New Horizons in Fluidization Engineering, Vancouver, Canada, 2007.

[42] A. Gómez-Barea, B. Leckner, Modeling of biomass gasification in fluidized bed, Prog. Energy Combust. Sci. 36 (4) (2010) 444–509.

[43] M. Jadidi, Introduction to Reactive Flows Modeling Using ANSYS FLUENT, Eddy-Dissipation Model (EDM), Reactive Flows, (2016).

[44] K. Kerst, C. Roloff, L.G.M.d. Souza, A. Bartz, A. Seidel-Morgenstern, D. Thévenin, G. Janiga, CFD-DEM simulations of a fluidized bed crystallizer, Chem. Eng. Sci. 165 (2017) 1–13.

[45] A.F. Kirkels, G.P.J. Verbong, Biomass gasification: still promising? A 30-year global overview, Renew. Sust. Energ. Rev. 15 (1) (2011) 471–481.

[46] A. Kumar, D.D. Jones, M.A. Hanna, Thermochemical biomass gasification: a review of the current status of the technology, Energies 2 (2009) 556–581.

[47] B.E. Launder, D.B. Spalding, The numerical computation of turbulent flows, Comput. Methods Appl. Mech. Eng. 3 (2) (1974) 269–289.

[48] Q. Li, M. Zhang, W. Zhong, X. Wang, R. Xiao, B. Jin, Simulation of coal gasification in a pressurized spout-fluid bed gasifier, Can. J. Chem. Eng. 87 (2009) 169–176.

[49] T. Li, S. Pannala, M. Shahnam, CFD simulations of circulating fluidized bed risers, part II, evaluation of differences between 2D and 3D simulations, Powder Technol. 265 (2014) 13–22.

[50] J. Li, X. Tian, B. Yang, Hydromechanical simulation of a bubbling fluidized bed using an extended bubble-based EMMS model, Powder Technol. 313 (2017) 369–381.

[51] H. Liu, A. Elkamel, A. Lohi, M. Biglari, Computational fluid dynamics modeling of biomass gasification in circulating fluidized-bed reactor using the Eulerian − Eulerian approach, Ind. Eng. Chem. Res. 52 (2013) 18162–18174.

[52] C. Loha, S. Gu, J. De Wilde, P. Mahanta, P.K. Chatterjee, Advances in mathematical modeling of fluidized bed gasification, Renew. Sust. Energ. Rev. 40 (2014) 688–715.

[53] X. Lu, T. Wang, Investigation of radiation models in entrained-flow coal gasification simulation, Int. J. Heat Mass Transf. 67 (2013) 377–392.

[54] J. Ma, S.E. Zitney, Computational fluid dynamic modeling of entrained-flow gasifiers with improved physical and chemical sub models, Energy Fuel 26 (12) (2012) 7195–7219.

[55] P. Mellin, E. Kantarelis, W. Yang, Computational fluid dynamics modeling of biomass fast pyrolysis in a fluidized bed reactor, using a comprehensive chemistry scheme, Fuel 117 (2014) 704–715.

[56] S. Murgia, M. Vascellari, G. Cau, Comprehensive CFD model of an air-blown coal-fired updraft gasifier, Fuel 101 (2012) 129–138.

[57] P. Nakod, CFD modeling and validation of oxy-fired and air-fired entrained flow gasifiers, Int. J. Chem. Phys. Sci. 2 (6) (2013) 28–40.

[58] P.M. Nakod, R.E. Shelke, A review of sub-models for computation fluid dynamics (CFD) modelling of clean coal technology, Int. J. Adv. Res. Phys. Sci. 1 (7) (2014) 22–34.

[59] N.P. Niemelä, H. Tolvanen, T. Saarinen, A. Leppänen, T. Joronen, CFD based reactivity parameter determination for biomass particles of multiple size ranges in high heating rate devolatilization, Energy 128 (2017) 676–687.

[60] P.A. Nikrityuk, T. Förster, B. Meyer, Modeling of gasifiers: overview of current developments, in: P.A. Nikrityuk, B. Meyer (Eds.), Gasification Processes: Modeling and Simulation, Wiley, 2014.

[61] T.K. Patra, P.N. Sheth, Biomass gasification models for downdraft gasifier: a state-of-the-art review, Renew. Sust. Energ. Rev. 50 (2015) 583–593.

[62] P. Pepiot, C.J. Dibble, T.D. Foust, Computational fluid dynamics modeling of biomass gasification and pyrolysis, in: M. Nimlos, M.F. Crowley (Eds.), Computational Modeling in Lignocellulosic Biofuel Production, American Chemical Society, Washington, DC, 2010.

[63] R.P. Ramachandran, M. Akbarzadeh, J. Paliwal, S. Cenkowski, Computational fluid dynamics in drying process modelling—a technical review, Food Bioprocess Technol. 11 (2) (2017) 271–292.

[64] A. Ramos, E. Monteiro, V. Silva, A. Rouboa, Co-gasification and recent developments on waste-to-energy conversion: a review, Renew. Sust. Energ. Rev. 81 (2018) 380–398.

[65] M. Rüdisüli, T.J. Schildhauer, S.M.A. Biollaz, J.R. Ommen, Scale-up of bubbling fluidized bed reactors—a review, Powder Technol. 217 (2012) 21–38.

[66] S.K. Sansaniwal, K. Pal, M.A. Rosen, S.K. Tyagi, Recent advances in the development of biomass gasification technology: a comprehensive review, Renew. Sust. Energ. Rev. 72 (2017) 363–384.

[67] S. Schulze, M. Kestel, P.A. Nikrityuk, D. Safronov, From detailed description of chemical reacting carbon particles to subgrid models for CFD, Oil Gas Sci. Technol. − Rev. d'IFP Energ. nouv. 68 (6) (2013) 1007–1026.

[68] J. Sehrawat, M. Patel, B. Kumar, Gaussian process regression to predict incipient motion of alluvial channel, in: K.N. Das, K. Deep, M. Pant, J.C. Bansal, A. Nagar (Eds.), Proceedings of Fourth International Conference on Soft Computing for Problem Solving, Advances in Intelligent Systems and Computing, Springer, India, 2015.

[69] A. Shehzad, M.J.K. Bashir, S. Sethupathi, System analysis for synthesis gas (syngas) production in Pakistan from municipal solid waste gasification using a circulating fluidized bed gasifier, Renew. Sust. Energ. Rev. 60 (2016) 1302–1311.

[70] A. Silaen, T. Wang, Effect of turbulence and devolatilization models on coal gasification simulation in an entrained-flow gasifier, Int. J. Heat Mass Transf. 53 (2010) 2074–2091.

[71] V.B. Silva, A. Rouboa, Using a two-stage equilibrium model to simulate oxygen air enriched gasification of pine biomass residues, Fuel Process. Technol. 109 (2013) 111–117.

[72] V. Silva, A. Rouboa, Predicting the syngas hydrogen composition by using a dual stage equilibrium model, Int. J. Hydrog. Energy 39 (1) (2014) 331–338.

[73] V. Silva, E. Monteiro, N. Couto, P. Brito, A. Rouboa, Analysis of syngas quality from Portuguese biomasses: an experimental and numerical study, Energy Fuel 28 (2014) 5766–5777.

[74] V. Silva, N. Couto, D. Eusébio, A. Rouboa, P. Brito, J. Cardoso, M. Trninic, Multi-stage optimization in a pilot scale gasification plant, Int. J. Hydrog. Energy 42 (37) (2017) 23878–23890.

[75] R.I. Singh, A. Brink, M. Hupa, CFD modeling to study fluidized bed combustion and gasification, Appl. Therm. Eng. 52 (2) (2013) 585–614.

[76] V. Sreedharan, CFD Analysis of Coal and Heavy Oil Gasification for Syngas Production, Aalborg University, 2012.

[77] M. Syamlal, L.A. Bisset, METC Gasifier Advanced Simulation (MGAS) Model, Technical Note, National Technical Information Services, Springfield, VA, 1992.

[78] D.E. Thompson, Verification, Validation, and Solution Quality in Computational Physics: CFD Methods Applied to Ice Sheet Physics, NASA Technical Reports Server, NASA, 2005.

[79] A. Vepsalainen, S. Shah, J. Ritvanen, T. Hyppanen, Bed Sherwood number in fluidised bed combustion by Eulerian CFD modelling, Chem. Eng. Sci. 93 (2013) 206–213.

[80] Y. Wang, L. Yan, CFD studies on biomass thermochemical conversion, Int. J. Mol. Sci. 9 (2008) 1108–1130.

[81] S. Werle, Numerical analysis of the combustible properties of sewage sludge gasification gas, Chem. Eng. Trans. 45 (2015) 1021–1026.

[82] Y.S. Wong, J.P.K. Seville, Single-particle motion and heat transfer in fluidized beds, AICHE J. 52 (12) (2006) 4099–4109.

[83] Y. Wu, P.J. Smith, J. Zhang, J.N. Thornock, G. Yue, Effects of turbulent mixing and controlling mechanisms in an entrained flow coal gasifier, Energy Fuel 24 (2010) 1170–1175.

[84] N. Xie, F. Battaglia, S. Pannala, Effects of using two- versus three-dimensional computational modeling of fluidized beds: Part II, budget analysis, Powder Technol. 182 (2008) 14–24.

[85] N. Xie, F. Battaglia, S. Pannala, Effects of using two- versus three-dimensional computational modeling of fluidized beds Part I, hydrodynamics, Powder Technol. 182 (2008) 1–13.

[86] L. Yu, J. Lu, X. Zhang, S. Zhang, Numerical simulation of the bubbling fluidized bed coal gasification by the kinetic theory of granular flow (KTGF), Fuel 86 (2007) 722–734.

[87] Y. Zhang, F. Lei, Y. Xiao, Computational fluid dynamics simulation and parametric study of coal gasification in a circulating fluidized bed reactor, Asia Pac. J. Chem. Eng. 10 (2) (2015) 307–317.

SECTION II

Combustion modeling

CHAPTER 3

Overview of biomass combustion modeling: Detailed analysis and case study

1. Introduction

Combustion still spearheads the incumbent technologies to turn fossil and nonconventional fuels into energy. About 70% of the world's energy supply comes from combustion processes [1], and despite the emergence of new technologies, all the forecasting scenarios ascribe a similar perspective in the future and even increasing the sharing regarding renewable sources. The combustion influence in critical sectors such as transportation, heating, and manufacturing processes demands a set of constant improvements on:
- handling with different fuels;
- getting superior efficiencies;
- complying with more restrictive environmental regulations.

Fossil fuels combustion dominates the international scene but the threatening views of a world moving to unprecedented global warming required the use of biomass and other wastes as a more sustainable source of energy. However, the use of such raw materials instead of coal, gasoline, or diesel raises a set of new technical challenges where computational science plays an indispensable role. The fundamental mechanisms of a combustion regime addressing such a kind of nonconventional fuels are not far away from the conventional fuels but also include a new set of chemical reactions, a collection of mixed fuels with demanding kinetic equations, and different ways of pollutant formation. Besides, other challenging questions as robust optimization, uncertainty quantification, and the design of new configurations assign a new set of undeniable problems to solve. All these issues are not fully understood and require in-depth experimental and numerical work. From an engineering perspective, the use of high-fidelity models avoids expensive laboratory tests and saves considerable time. Despite a large number of papers being found in the literature [2–4], such works are often hard to reproduce, and most engineers and researchers struggle with how to implement and develop the numerical models.

Computational Fluid Dynamics Applied to Waste-to-Energy Processes
https://doi.org/10.1016/B978-0-12-817540-8.00003-0

© 2020 Elsevier Inc.
All rights reserved.

On the other hand, there are a bunch of questions regarding how to accommodate the kinetic mechanisms, the fuel features, and simplifications to take. Since a lot of species and even more reactions are involved, how do they cope with them? What is the ideal number to include in the simulations? Which are the methods to accelerate the solver converging using detailed mechanisms? How do they include more than one fuel? Which approach is better, Eulerian-Lagrangian or Eulerian-Eulerian?

The next sections intend to provide critical elements for such questions and help the user to decide under different problems. This chapter aims at presenting a step-by-step guide of how to implement a DPM to simulate the combustion of Eucalyptus residues in a pilot-scale fluidized bed combustor. One first describes the experimental setup and the substrate properties, and then explores the different software options to disclose the nature of the interactions between the chemistry sets, fluid dynamics, the mass, momentum, and heat exchange between phases, as well as the dynamics of modeling the pollutant formation mechanisms with the focus in NO_x particles.

2. Experimental setup

The purpose of this tutorial is to lead the user into modeling and analyzing the parameters of interest over the fluidization process during biomass combustion by employing a 3D model. The reactor considered in this tutorial is presented in Fig. 1. The unit refers to a

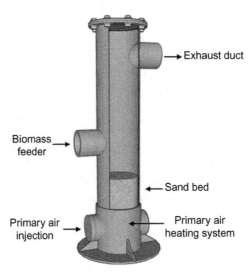

Fig. 1 Schematics of the pilot-scale bubbling fluidized bed reactor.

Table 1 Proximate and ultimate analysis.

Elemental analysis (wt%, dry basis)	Eucalyptus
Ash	2.87
C	45.85
H	6.13
N	0.35
S	$\leq$100 (ppm)
O (by difference)	44.80
Proximal analysis (wt%, wet basis)	
Moisture	11.8
Ash	2.6
Volatile matter	71.0
Fixed carbon	14.6

75 kW$_{th}$ pilot-scale bubbling fluidized bed reactor located at the University of Aveiro, Portugal. The reactor holds a vessel with an internal diameter of 0.25 and 2.3 m height. The bottom bed carries a static height of 0.23 m comprising 17 kg of quartz sand with particles ranging between 355 and 710 μm. Eucalyptus wood (*Eucalyptus globulus*) is employed as biomass during the simulation. Dry atmospheric air is used as an oxidizing agent, being fed into the reactor throughout the bottom of the sand bed. The produced gas (syngas) leaves the reactor throughout an outlet placed at the top right corner of the geometry. Simulations were calculated using a time step size of 0.001 s, for a total number of 3000 time steps (3 s).

The proximate and ultimate analysis is crucial information to set the global combustion equation. Table 1 shows the analyses for Eucalyptus residues.

3. Creating the geometry

While creating the geometry, one is affecting the mathematical model since it defines the domain in which the governing equations are defined. Moreover, boundary conditions are also defined at the edges of the domain. In this tutorial, the geometry is created from scratch in ANSYS DesignModeler. The user must first make an effort in trying to simplify or remove unnecessary features. A simplified 3D representation of the reactor beforehand is to draw a cylinder with the same dimensions.

3.1 Start ANSYS DesignModeler.

3.2 Set the units in ANSYS DesignModeler to Meter.

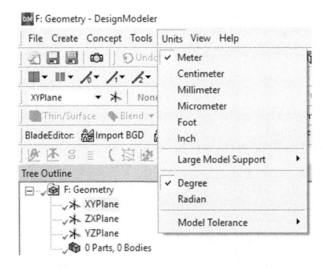

3.3 Select the XYPlane from the Tree Outline and then click on the New Sketch button.

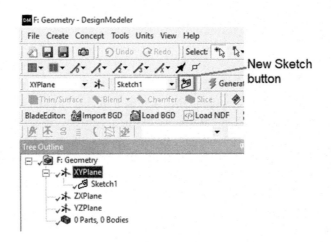

3.4 On the Sketching tab, select Circle from the Draw toolbox and in Graphics draw the circle by clicking on the origin of the plane.

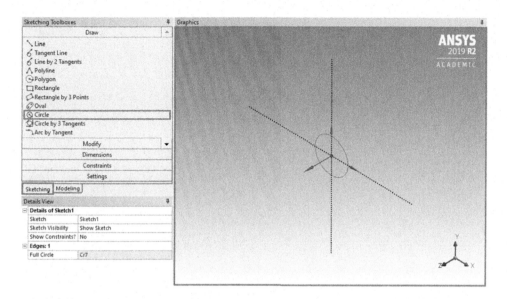

3.5 Input the fluidized bed reactor diameter dimensions for the Circle by clicking in the Dimensions tab and then set the Circle option, placing 0.25 m for width.

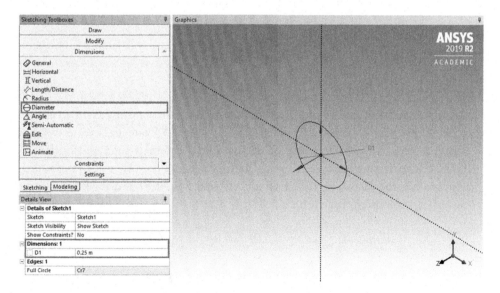

3.6 To create the 3D cylinder, click on Extrude, select the Circle drawn (Sketch1) for Geometry and press Apply. In FD1, Depth (>0), set the reactor's height to 2.3 m, Direction as Normal and press Generate.

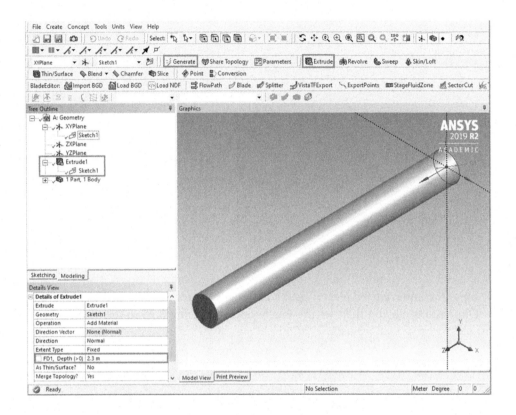

3.7 Having created the main cylinder, it is now time to create opposing smaller cylinders to depict the biomass inlet and outlet. Click on the ZXPlane and create two new Sketches (Sketch2 and Sketch3) for each cylinder, biomass inlet and outlet. On the ZXPlane, draw two circles, one at the bottom and the other at the top of the cylinder. Set the diameter (0.11 m for both) and height (0.3 m for inlet and 2 m for outlet) dimensions for each circle.

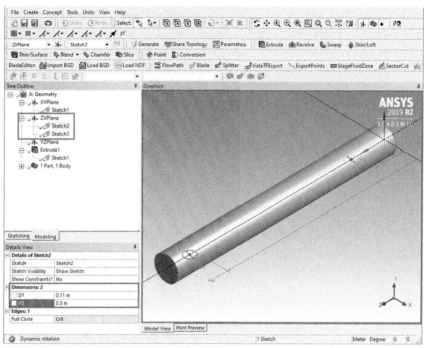

3.8 Click on Extrude to create the two 3D cylinders to stand as biomass inlet and outlet. For both circles, set FD1, Depth (>0) as 0.2 m for length, Direction as Normal for Extrude2 and Reversed for Extrude3, and Generate.

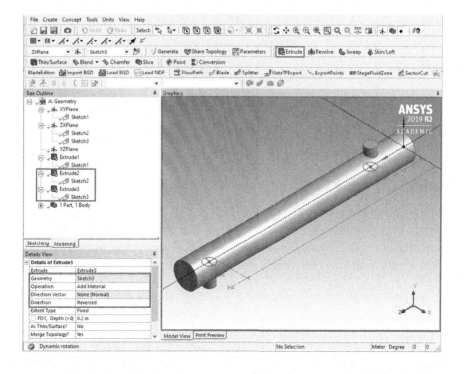

3.9 The 3D simplified reactor geometry is now fully created. On the menu toolbar, select Selection Filter: Faces (for geometry surface), click on the surfaces of the geometry, and as the boundaries turn green, right-click on the boundary and press Named Selection. Set the inlet, outlet, biomass_inlet, wall, and the whole geometry surface as fluid. Following all these steps, the geometry is now created; the user may now Save the Project if using ANSYS Workbench, or export the geometry file.

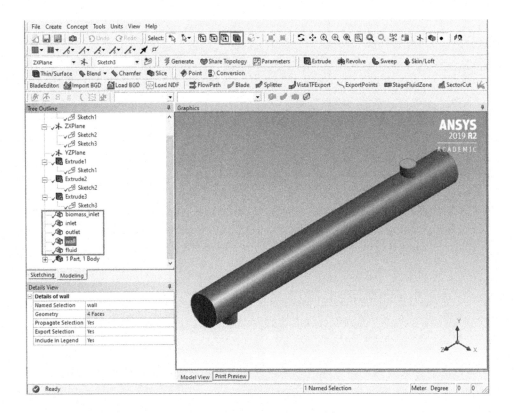

4. Creating the mesh

Numerical methods require the geometry to be split into discrete cells, usually referred to as elements. This process is known as meshing. The ability of numerical methods to accurately predict results relies upon the mesh quality. The optimal mesh is the one that maximizes accuracy and also minimizes the solver run time.

4.1 Start ANSYS Meshing and import geometry.

4.2 Click on Mesh in the Tree Outline to show the Details of "Mesh," and make sure the Physics Preference is set to CFD and the Solver Preference is set to Fluent.

4.3 Expand Sizing toolbox and confirm that Capture Curvature and Proximity are on, then expand the Quality toolbox and turn Smoothing to High. You may now Generate the Mesh.

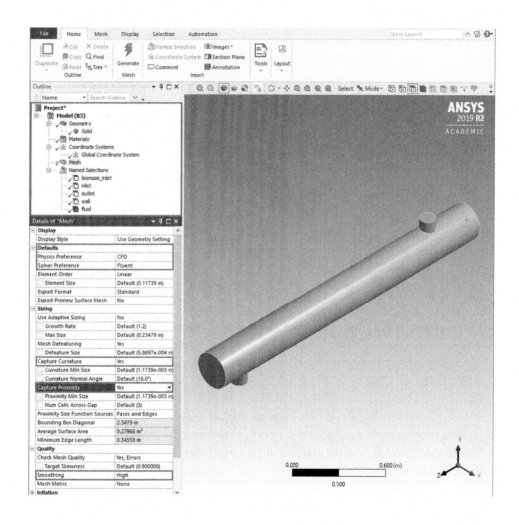

4.4 Click on the Body bottom and select the whole geometry, then click on Mesh tab and select Sizing from the drop-down list, and press Apply to create a Body Sizing feature. In the Details of "Body Sizing," set the element size as 0.0181 m and Generate Mesh.

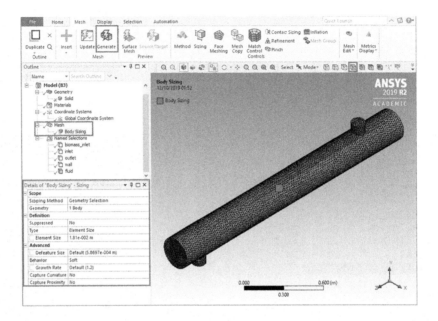

4.5 Having created the mesh, one may check the Statistics for the number of Nodes and Elements contained in the mesh. A number placed around 167,000 elements is considered sufficient for the study in hand. To check the quality of the mesh, select Element Quality in Mesh Metric from the Quality drop list; an Element Metrics will be made available in the Mesh Metrics. Element quality ranges from 0 to 1, in which higher values indicate higher element quality. If the metrics show a proper mesh quality, the user may now Save the Project if using ANSYS Workbench, or file Export and specify Fluent Input File (.msh) if using standalone Fluent.

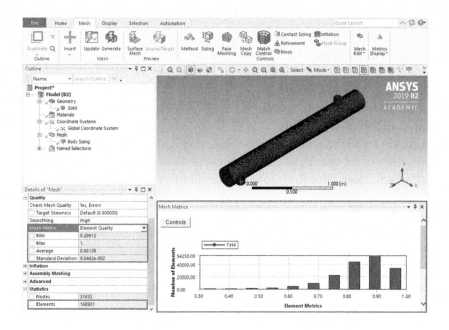

5. Model considerations and implementation

Having created a faithful representation of the domain and discretized it into a finite set of control volumes, it is now required that the users configure both physics and solver values based on their knowledge of the fluid application. This setup stage is set within the ANSYS Fluent framework, requiring the general solver settings, the definition of physical and chemical mathematical models, material properties, phases creation and interactions, operating cell zone and boundary conditions.

The implementation of a combustion model is always complex and should follow a pathway for a numerically stable simulation. In fact, the introduction of all the underpinning mechanisms behind the combustion process simultaneously will certainly lead to difficulties in order to get a converging solution. Any mathematical problem will hit the right solution easier when departing from a good initial point. The converging procedure will run smoothly following the steps below:

- obtain a first solution implementing a cold flow analysis without enabling reactions, radiation, and fluid–particle interactions;
- patch higher temperatures (1500–2000 K) in the flame region;
- enable the Discrete-Phase Method and run 1 iteration for ignition;
- run the reactive multiphase flow;
- implement the radiation phenomenon;
- run the particle radiation effects.

This procedure can be shortened if the user has previous knowledge about the best initial conditions. Anyway, the problem solving in several steps will always smooth the solution accomplishment. **Then, the first goal for simulation should target the establishment of a stable cold flow (with no reactions, no particles, no radiation, only flow and nonreacting heat transfer).**

5.1 Start ANSYS Fluent and ensure that the options Display Mesh After Reading, Workbench Color Scheme, Double Precision, and Serial Processing in the ANSYS Fluent Launcher are enabled. In general, single precision should be eschewed when there are significant differences in the geometry components length.

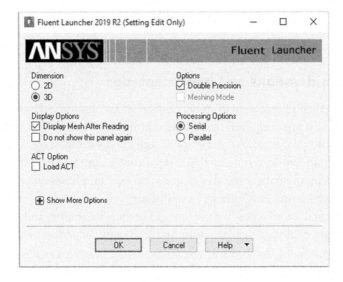

5.2 Select the Setup drop list and start by setting the General options. Click on Check and Report Quality to inspect the Mesh Quality. In Solver settings, enable Type as Pressure-Based, Time as Transient, and Velocity Formulation as Absolute. Enable the Gravity option in the Z direction and input the gravitational acceleration, -9.81 m/s^2.

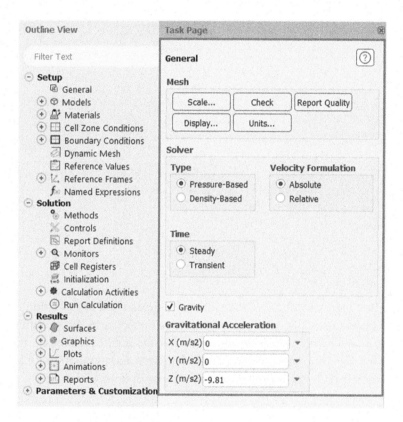

The user should tap the Energy Model, the Viscous Model as SST k-omega (2 eqn), and enable Species Transport.

5.3 Set the Boundary conditions, define inlet as velocity–inlet, interior–fluid as interior, outlet as pressure–outlet, and wall as wall. For the Velocity Inlet, define Velocity Magnitude as 0.35 m/s, Turbulent Intensity 5%, Hydraulic Diameter 0.13 m (diameter for the distributor plate). Click the Thermal tab and input 500 K. Click Species tab and define 0.23 for O_2. Retain all remaining options as default.

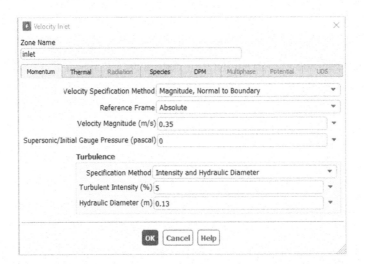

5.4 As for the Pressure Outlet, ensure Backflow Direction is Normal to Boundary, select Specification Method as Intensity and Viscosity Ratio, and set 5% for Backflow Turbulent Intensity and 10 for Backflow Turbulent Viscosity Ratio. Click the Thermal tab and input 300 K. Click Species tab and ensure all species mass fractions are null. Boundary conditions for the wall may be left as default, no actions required.

5.5 Define the Solution Methods, set Simple for Scheme and for Spatial Discretization, Gradient: Least-Squares Cell Based; Pressure: PRESTO!; Momentum: Second Order Upwind; Turbulent Kinetic Energy: First Order Upwind; and Specific Dissipation Rate: First Order Upwind.

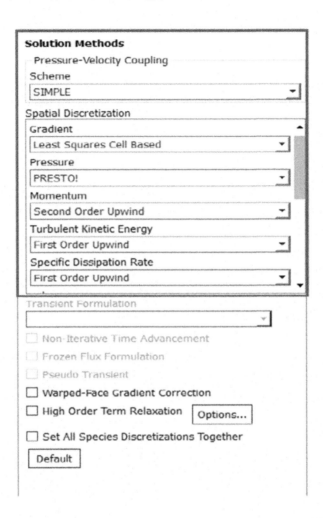

5.6 Set the Under-Relaxation Factors, Pressure: 0.3; Density: 1; Body Forces: 1; Momentum: 0.7; Turbulent Kinetic Energy: 0.6; Specific Dissipation Rate: 0.8; Turbulent Viscosity: 0.8; and Vol: 0.8. Then, reorder the mesh and run at 500 or more iterations until convergence. Save the cold flow case and data files.

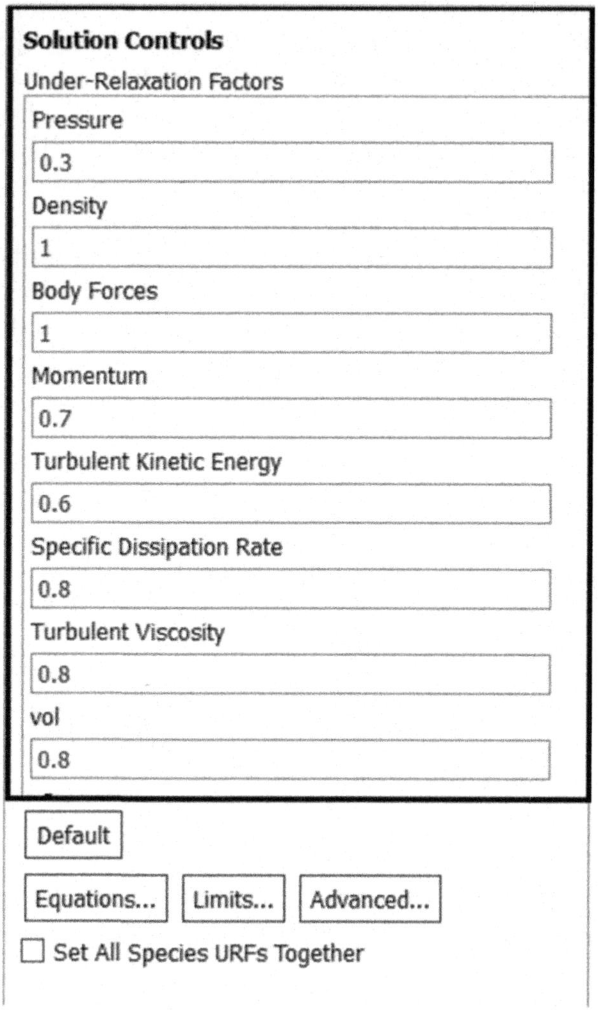

Solution Controls

Under-Relaxation Factors

Pressure

| 0.3 |

Density

| 1 |

Body Forces

| 1 |

Momentum

| 0.7 |

Turbulent Kinetic Energy

| 0.6 |

Specific Dissipation Rate

| 0.8 |

Turbulent Viscosity

| 0.8 |

vol

| 0.8 |

| Default |

| Equations... | | Limits... | | Advanced... |

☐ Set All Species URFs Together

The simulation should proceed with the **Reactive Multiphase Flow** (with no radiation as the first step). So far, the user has only computed the flow field and considered one continuous phase. The user should now set up the fuels, mixtures, and reactions.

5.7 In the Species Model tab, click on Edit right after coal-hv-volatiles-air. The Edit Material tab will be made available. Fluent has some inbuilt mixtures such as coal, lignite, or wood. The other option is to build new material. Depending on the fuel, this is the most likely option. In the case of using biomass, the user can define a new **biomass-volatiles–air mixture** but needs to proceed with further calculations.

The process is explained in detail along the next lines. By clicking on Edit, the user can observe the gas components comprising the mixture. Some species like CO are not present and must be included manually. Before advancing, the user should previously identify all the necessary species for the problem. Fluent has solid species but they are associated with shell materials like aluminum. Char, which is crucial for the combustion problem design, is here included in the fluids box as carbon <solid> and acts as a pseudo–fluid due to mathematical issues. When the user opens the mixture, the carbon <solid> should fill the solid section. For mathematical reasons, nitrogen should be the last specie under the fluid window and they should be ordered by their predictable amounts. In fact, if the user finds that the last specie in the Selected Species list is not the most abundant specie (as it should be), he/she will need to rearrange the species to obtain the proper order.

Select Edit in Reaction and define the Species, Stoichiometric Coefficient, Rate Exponent, and the Arrhenius Rate. Fluent provides the coal calculator and the corresponding combustion equation for any type of coal based on the elementary and proximal analysis. Most of the time, the user's combustion problem will include a fuel that is not coal or even a pure specie (e.g., biomass or municipal solid waste). In such cases, the user needs to get available the heating value and the elemental composition of this new fuel, and then define an equivalent fuel specie and an equivalent heat of formation. Regarding the elemental composition, the user should divide each atom percentage by the corresponding atomic weight, and a new "fuel" molecular formula is created (e.g., $C_xH_yO_z$). For instance, if the elemental analysis shows 50% C, 6% H, and 44% O, the molecular formula should be $C_{4.17}H_6O_{2.75}$. Then, the user can assign a new molecular weight to this new "fuel" as follows: $x*C$ atomic weight $+ y*H$ atomic weight $+ z*O$ atomic weight. In this case, the new molecular weight would be 100.04 kg/kgmol. The next step involves the combustion equation where the products are CO_2 and H_2O. To set the appropriate stoichiometric coefficients, the user should balance the exact same number of the atoms of each element on each side. Depending on the fuel, the math could be extremely easy or for more complex calculations the excel solver will be a great help. In this case, the chemical reaction would be as follows:

$$C_{4.17}H_6O_{2.75} + 4.295\,O_2 \rightarrow 4.17\,CO_2 + 3\,H_2O \tag{1}$$

With this information, the user can now set the global equation as well as the equivalent heat of formation.

In the present case, for Eucalyptus combustion, the global equation is

$$C_{3.82}H_{6.13}O_{2.8} + 3.9525\,O_2 \rightarrow 3.82\,CO_2 + 3.065\,H_2O \tag{2}$$

With the global equation, the user can proceed to the volatiles equations:

$$vol + O_2 = CO + H_2O \tag{3}$$

To determine the stoichiometric coefficients, the user needs the proximate analysis and collects information about volatiles distribution. It is possible to get this last piece of information as follows:
- find a volatile distribution in the open literature for the fuel under study;
- determine experimentally the volatile distribution (This was our case, and due to a nondisclosure agreement, our results are not yet available for the general public);
- consider ratios between gaseous species (e.g., $CO/CO_2 = 0.7$);
- volatile breakup approach.

In this last approach, Nakod [5] solves a system of simultaneous equations for the mass balance of each element in the volatiles equation. However, the author finds seven unknown quantities to be evaluated from five equations. The solution to these equations can be obtained by making two suitable assumptions (the consideration of ratios between gaseous species could be extremely helpful here).

The reader has now a set of tools to select and implement in each case. The figure below shows the final volatile equation considering the proximate analysis and the obtained results considering the experimental volatiles distribution:

$$vol + 2.08\,O_2 = 1.90\,CO + 2.38\,H_2O \tag{4}$$

Besides the volatile equation, the user can add a second equation for the oxidation of CO to CO_2:

$$CO + 0.5\,O_2 = CO_2 \tag{5}$$

The user may optionally set the remaining chemical equations to solve the multiple char reactions, O_2 (Eq. 6), CO_2 char (Eq. 7), H_2 (Eq. 8), and H_2O combustion (Eq. 9), as follows:

$$C(s) + 0.5O_2 = CO \tag{6}$$

$$C(s) + CO_2 = 2CO \tag{7}$$

$$C(s) + H_2O = H_2 + CO \tag{8}$$

$$H_2 + 0.5O_2 = H_2O \tag{9}$$

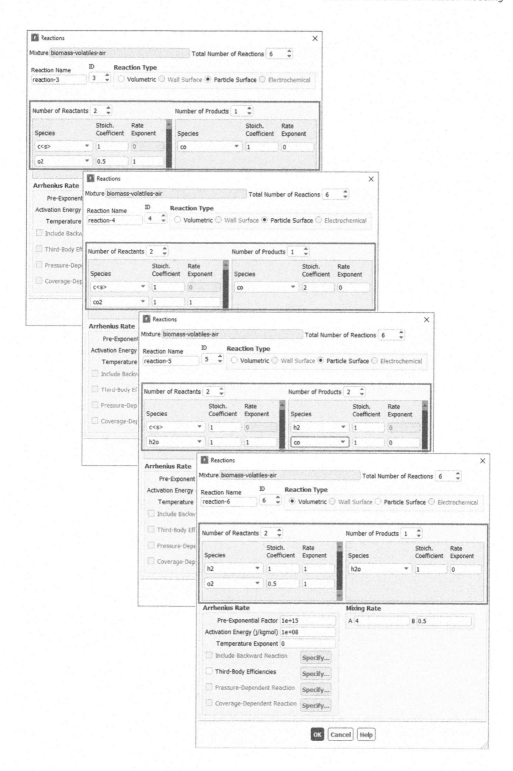

5.8 Select Edit in Mechanism and ensure that both chemical reactions (Eqs. 1 and 2) are properly selected.

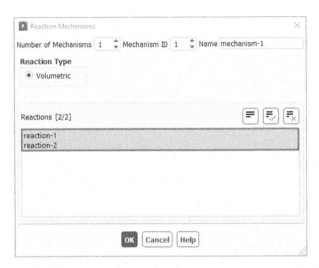

5.9 Edit the biomass Combusting Particle by adding the Eucalyptus wood properties, set 478 for Density (kg/m^3) and 1374 for Cp (Specific Heat) (J/kg-K), and leave all remaining options as default.

After establishing a stable cold flow, defining the chemical reactions and boundary conditions, the user should focus attention on the **Ignition** process. The fuel was already defined but it is still not available to react. To proceed with the reaction, the user needs to inject fuel particles and simultaneously **provide the required energy to trigger the volatilization**. A good way to lever the reactions goes through the **patching of some product species or high temperatures,** meaning by this that the user could initiate the reaction with a predefined amount of a product (e.g., to set the CO_2 fraction as 0.1).

5.10 Set up the Discrete-Phase Model, enable Interaction with Continuous Phase, set DPM Iteration Interval to 1, and set the Tracking Parameters by inputting 40000 for Max. Number of Steps and 0.0025 m for Length Scale.

5.11 A set of particle injections properties can be created starting from a specified zone, biomass_inlet in this particular case. The biomass particles once injected into the reactor take a short distance before releasing volatiles. To set this, click on Injections; here, the user may read, edit, or create an injection file. Set the injection details for injection-0; this will open the Set Injection Properties window. Define Injection Type as surface, select Combusting for Particle Type, hv_vol for Devolatilizing Species, O_2 for Oxidizing Species, and CO_2 for Product Species. In the event of building a new injection without an injection file, the user must set the velocity components, the diameter, the temperature, and mass flow. In this case, the Point Properties were designated as follows: 23.11 for Z-Velocity, 1e-6 for Diameter, 343 K for Temperature, and Total Flow Rate as 1e-20.

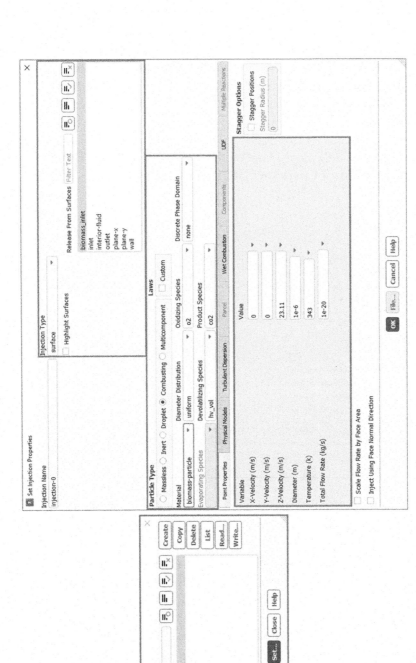

5.12 Define an adaption register for the bed height of the fluidized bed. In the Adapt tab, click Refine/Coarsening > Cell Registers > New > Region and set the Region Adaption, X Max 0.25 m for the bed width and Y Max 0.23 m for the bed height, click Save/Display to refine cells and Close.

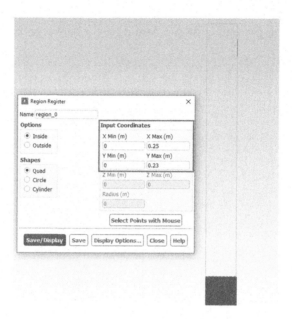

5.13 Initialize the solution, set 84 for Turbulent Kinetic Energy (m²/s²); 5550 for Specific Dissipation Rate (1/s); 0.23 for O_2; and 2000 K for temperature. Press Initialize, then Patch the temperature and mass fractions to the adapted region marked in the fluidized bed reactor. In Patch, select Variable > Temperature > enter 2000 for Value > select the Registers to Patch and press Patch. Repeat the same steps adding 0.01 for both CO_2 and H_2O mass fractions, patch each gas specie individually, and close.

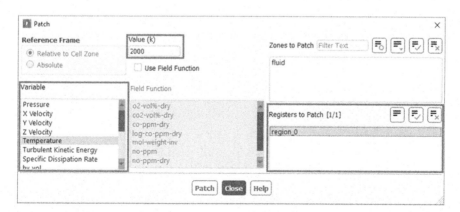

5.14 The user may now run the calculation. First, in the Solution Controls, set the Discrete–Phase Sources Under-Relaxation Factor to 1. During this stage, one must first perform a single iteration run to proceed with the ignition and subsequently save Case and Data file. This is a very important step and many users miss it by moving directly to a large number of iterations. Then, in the Discrete-Phase Model parameters, one must reset the DPM Iteration interval to 50, further shift the Discrete-Phase Sources Under-Relaxation Factor to 0.1, perform additional 350 iterations, and save Case and Data file. At this point, the user must assess whether this number of iterations is sufficient as this procedure varies according to the convergence criteria being used.

After getting the convergence with the Reactive Multiphase Flow, it is now time to include **Radiation**. In combustion processes, radiation plays a significant role in the heat transfer analysis. Fluent disposes several Radiation models to describe heat transfer simulations. Different radiation models provide different solutions; therefore, some radiation models may be more appropriate to solve certain problems than others. Moreover, these differ in their accuracy and CPU requirements. In this tutorial, the Discrete Ordinates (DO) model is employed and grants an accurate solution for this problem. Also, this model is commonly applied to combustion cases. To engage this model, search within the Models drop-down list and select Radiation to open the Radiation Model dialog box. Enable the Discrete Ordinates Model, set 4 for both Theta and Phi Divisions and 3 for Theta Pixels and Phi pixels, set 1 Energy Iterations per Radiation Iteration and run 550 iterations to converge the model. Save Case and Data file. Again, the user must determine if the number of iterations considered satisfies the converge criteria of the solution. Monitoring the residuals during the calculation is a good practice as they stand as a useful indicator of the iterative converge of the solution. Throughout this iterative process, the residuals are expected to progressively decay to smaller values up until they get leveled and substantial changes stop occurring.

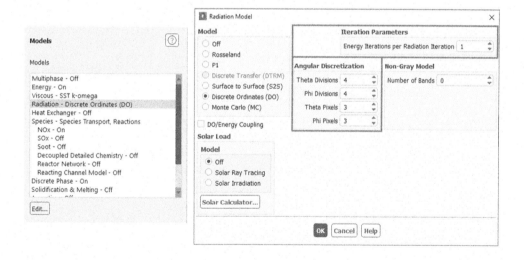

If the user needs a more accurate simulation, the inclusion of the **Particle Radiation** effect is also very important. In the Discrete-Phase Model dialog box, enable Particle Radiation Interaction under Physical Models tab. Head to Materials > Combusting Particle > biomass-hv and enter 773 K for Vaporization Temperature; ensure that the Particle Scattering Factor is set to 0.15. Run 2000 iterations to achieve convergence and save Case and Data file.

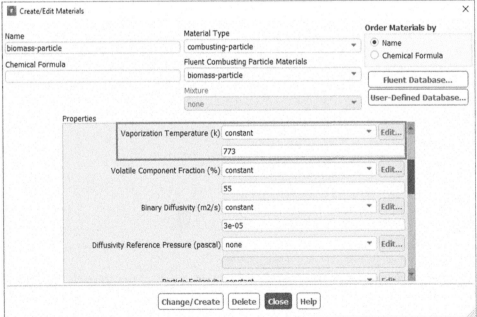

6. Results display

After the solution be converged, the next step is to process the obtained results. Postprocessing options can be directly applied in the ANSYS Fluent postprocessing or in the ANSYS CFD-Post. Here, the results attained will be presented in greater depth in the postprocessing chapter, yet an example of the possible results to be drawn from this sort of simulation is provided.

Regarding the convergence criterion, residuals history hands over meaningful insights over convergence. Generally, residuals decrease and stabilization are good signs of proper solution convergence. Nevertheless, these alone should not be considered as strict convergence criterion.

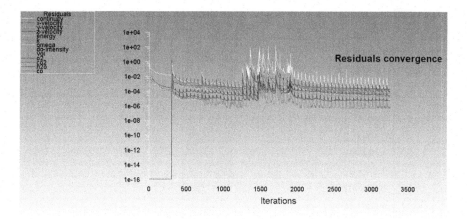

Therefore, apart from residuals monitoring, Flux Reports provide additional important convergence indicators. To employ this, go to Results > Reports > Fluxes opening the Flux Reports dialog box. Here, enable Mass Flow Rate from Options, select biomass_inlet and outlet from Boundaries and Compute. The net flux imbalance, shown in Net Results, should end in a result inferior to 1% of the smallest flux through the smallest inlet/outlet result. The DPM Mass Source is positive whenever the particle is a source of mass in the continuous phase.

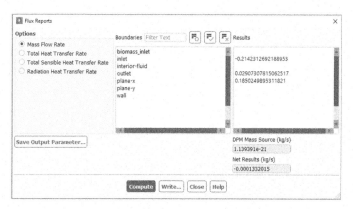

– Plot contour of temperature within the reactor on surface $x = 0$.

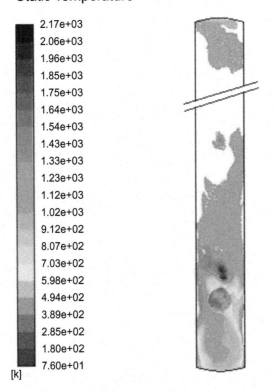

– Plot contour of O_2 mass fraction within the reactor on surface $x = 0$.

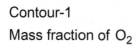

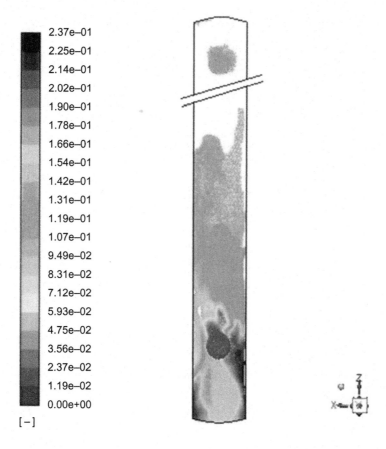

7. NO$_x$ formation

Fig. 2 provides the comparison focused on some of the most important air pollutants, the CO_2 generated per kWh, NO_x, sulfur dioxide (SO_2), and particulates generated per g/ton of processed fuel for both gasification and combustion conversion power plants. Undoubtedly, gasification technology is less adverse to the environment showing reduced emissions per unit of generated power as compared to combustion systems. Emissions control is made simpler in gasification than in combustion since the produced gas in gasification comes at higher temperatures and pressures as compared to the combustion's exhaust gases,

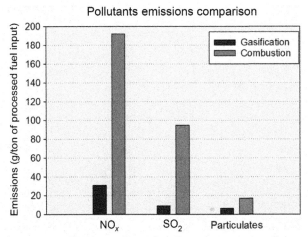

Fig. 2 Comparison between gasification and combustion processes air pollutants [6].

promoting the easier removal of pollutants and traces of contaminants. In the particular case of NO_x emissions, they are known for their harmful effects on the environment, such as in the air quality by promoting photochemical reactions in the atmosphere and also in human health [7]. NO_x emissions carry on negative environmental effects and must be reliably predicted and minimized to prevent significant environmental and economic costs. ANSYS Fluent provides the capability to model the NO_x formation and consumption in combustion systems. It includes a set of embedded models developed at the Department of Fuel and Energy at the University of Leeds in England, as from the available literature. The NO_x prediction comes up from the resolution of the transport equation for NO_x concentration, and when information is available, it is possible to solve transport equations for intermediate species such as HCN, NH_3, and N_2O. The reader must take into consideration the fact that the transport equations are only solved after processing the entire combustion solution, meaning that the NO_x modeling accuracy is intimately linked with the previous use of an effective combustion model. The computation of NO_x transport equations in a postprocessing stage takes advantage of the tiny values of the mass fractions of these species, then exerting small or any influence on all other flow fields and chemical reactions. Such a feature brings the following additional shortcomings:

- the user can only use the NO_x model within the steady-state solver;
- it is hard to revert the initial settings of a case file previously set up for the decoupled detailed chemistry NO_x model. However, the user could overtake this issue by saving the files before decoupling the NO_x model.

Having said that, it is obvious that the reader can only expect the results to be as accurate as the selected inputs, chemical and physical models. Also, the user must understand that is almost impossible to achieve accurate absolute values, but the ability to predict the NO_x trends as a function of different operating conditions prevails. This goes in line with the effective power of CFD simulations, the effective prediction of parameter effects allowing a significant save of time and money that otherwise would be necessary for performing countless laboratory experiments.

To engage NO_x modeling in ANSYS Fluent, Import the Case file into Fluent, go to Models drop-down list and below Species select NO_x model. From the Pathways list, enable Thermal NO_x, Prompt NO_x, and Fuel NO_x. In Formation Model Parameters under the Thermal tab, select partial-equilibrium for both [O] and [OH] Model. In Prompt tab, select vol from the Fuel Species list, and input 2.8 for Fuel Carbon Number and 0.685 for Equivalence Ratio. In the Formation Model Parameters within on Fuel tab, the user may set the Fuel parameters. These parameters are highly variable according to the Fuel one wishes to model. Click on Turbulence Interaction Mode tab, select temperature from the drop-down list, and enter 20 for Beta PDF Points. Click on Apply and Close the NO_x model dialog box. Before resuming with the calculation, go to the Solution > Controls > Equations and deselect all equations except the pollutants one wishes to model. Then, in Monitors > Residuals, ensure that the convergence criteria for the pollutant's equations are set to 1e-06, run 200 iterations, and save the Case and Data file. Ensure that 200 iterations satisfy the converge criteria of the solution, or if additional iterations are required.

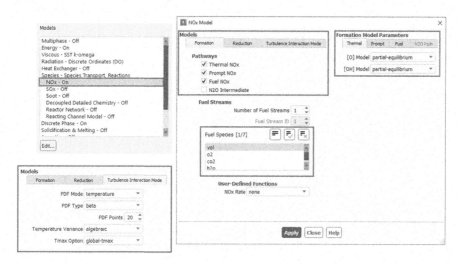

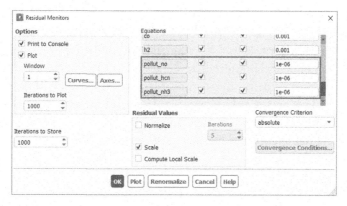

– Plot the contour of the pollutant NO_x mass fraction on surface $x = 0$.

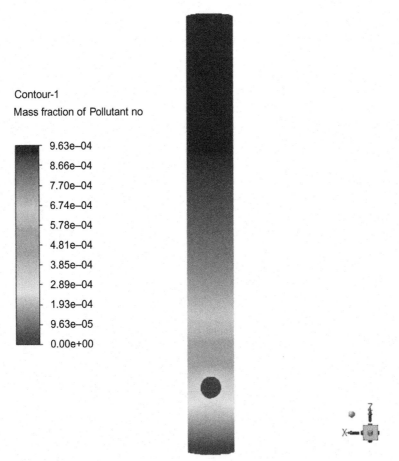

The NO_x mass fraction formation range goes in hand with the trends found in the literature for biomass combustion [8, 9].

8. Final remarks

Conclusively, this tutorial presents a 3D approach to simulate the combustion of Eucalyptus biomass in a pilot-scale fluidized bed reactor. As shown, during the process, the user is allowed to set the chemical equations mechanism; here, a simpler approach has been proposed. However, the user may set more complex chemical model approaches to better suit the problem in hand by importing chemical equations mechanism files into the ANSYS Fluent framework. Additionally, user-defined functions (UDF) routines can be implemented to improve the model, allowing to alter the general correlations within the ANSYS Fluent database so as to fit the intended modeling needs and to better agree with our experimental setup. Such routines were left out of consideration in this tutorial guide, as their application is thoroughly discussed in other chapters. Quantifying the pollutants is a meaningful step to achieve an accurate and complete combustion solution. The ANSYS Fluent NO_x model provides the capability to model thermal, prompt, and fuel NO_x formation and NO_x consumption due to reburning in the combustion system. Beyond NO_x modeling, ANSYS Fluent also provides additional specific pollutant models such as SO_x and soot, with fixed chemical mechanisms.

References

[1] International Energy Agency (IEA), Global energy and CO_2 status report 2018, 2019.
[2] V. Silva, E. Monteiro, N. Couto, P. Brito, A. Rouboa, Analysis of syngas quality from Portuguese biomasses: an experimental and numerical study, Energy Fuel 28 (2014) 5766–5777.
[3] J. Cardoso, V. Silva, D. Eusébio, P. Brito, R.M. Boloy, L. Tarelho, et al., Comparative 2D and 3D analysis on the hydrodynamics behaviour during biomass gasification in a pilot-scale fluidized bed reactor, Renew. Energy 131 (2019) 713–729.
[4] N. Couto, V. Silva, J. Cardoso, A. Rouboa, 2nd law analysis of Portuguese municipal solid waste gasification using CO_2/air mixtures, J. CO_2 Util. 20 (2017) 347–356.
[5] P.M. Nakod, A review of sub-models for computation fluid dynamics (CFD) modelling of clean coal technology, Int. J. Adv. Res. Phys. Sci. (IJARPS) 1 (2014) 22–34.
[6] J. Cardoso, V. Silva, D. Eusébio, Techno-economic analysis of a biomass gasification power plant dealing with forestry residues blends for electricity production in Portugal, J. Clean. Prod. 212 (2019) 741–753.
[7] N.A. Abaimov, A.F. Ryzhkov, Development of a model of entrained flow coal gasification and study of aerodynamic mechanisms of action on gasifier operation, Therm. Eng. 62 (2015) 767–772.
[8] C. Ghenai, I. Janajreh, Combustion of biomass and waste-based syngas fuels, in: Industrial Waste & Wastewater Treatment & Valorisation, Athens, Greece, 2015.
[9] B.A. Albrecht, R.J.M. Bastiaans, J.A. van Oijen, L.P.H. de Goey, NO_x Emissions Modeling in Biomass Combustion Grate Furnaces, Eindhoven University of Technology, 2006.

CHAPTER 4

Overview of biomass gasification modeling: Detailed analysis and case study

1. Introduction

Climate change has made its way as one of the most pressing problems of our time threatening the regular balance of the planet and with it the livelihood of billions of people. Presently, fossil fuels maintain a dominant position in meeting the world's energetic demands with renewable sources still making a small contribution. Responding to climate change requires a set of strategies that must be unfolded, out of these surveying biomass-derived energies may come as a promising solution to fight back the established fossil fuel dependency [1].

Gasification can be best described as a process of thermochemical conversion of carbonaceous raw materials into a combustible gas through the controlled delivery of an oxidizing agent (air, vapor, oxygen, or carbon dioxide). It allows converting feedstocks into higher heating value fuels by feeding the oxidizing agent below the stoichiometric values to prevent combustion and increase the efficiency of the process through oxidation. Fluidized bed technology is commonly applied to convert biomass into energy through gasification, offering several attractive features such as easy construction and operation; flexibility regarding multi characteristic fuel input; improved mixing capacities; uniform temperature distribution; less slagging and fouling problems; high heat transfer rates; high thermal efficiency; improved gas yields; and energy conversion [2].

Mathematical models have been broadly applied to solve some of the most intricate phenomena delivering a simplified representation of the real world. The same applies to gasification processes with mathematical modeling granting different approaches to interpret it. Concerning time dependence, these can be divided into kinetic models and kinetic-free models, if reaction kinetics and transport phenomena during the gasification process are to be considered or not, respectively [3]. As for the reactor geometry, gasification can be parted into zero-dimensional, one-dimensional, two-dimensional, and three-dimensional numerical models [4]. Among the several numerical approaches, advanced numerical methods such as computational fluid dynamics (CFD) exhibit utmost importance. CFD is a powerful tool to predict and model gasification experiments in fluidized bed gasifiers. It stands as a cost-effective option to explore different

Computational Fluid Dynamics Applied to Waste-to-Energy Processes
https://doi.org/10.1016/B978-0-12-817540-8.00004-2
© 2020 Elsevier Inc.
All rights reserved.

configurations and operating conditions at any scale allowing to identify the most stable set of conditions to operate the system [5]. Its framework considers energy flow, momentum, hydrodynamics, conservation of mass and turbulence in a defined region, focusing on the mixing of both solid and gas phases through the use of advanced numerical methods, and acknowledges heterogeneous chemistry of biomass gasification alongside with devolatilization, char combustion, and gas–phase chemistry [6, 7]. Unquestionably, gasification is extremely complex to describe mathematically, therefore one must perceive a deep knowledge of all relevant phenomena involved so to appropriately create a model capable of providing an accurate prediction of all the entangled phenomena occurring during the gasification process.

In this sense, this chapter aids researchers by handing out the numerical tools available within the ANSYS Fluent framework to successfully implement a 2D Eulerian-Eulerian model to simulate the biomass gasification in a pilot-scale bubbling fluidized bed reactor. At the end of this chapter, the user is entrusted with the required set of skills to perform CFD simulations of thermochemical processes while grasping design techniques, testing, and analyzing results.

2. Problem specification

This tutorial chapter guides the user to model and analyze the parameters of interest over the fluidization process during biomass gasification by employing a 2D model. Fig. 1

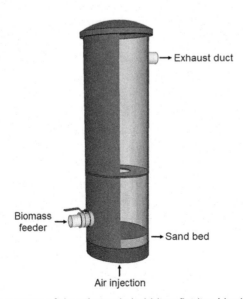

Fig. 1 Schematic representations of the pilot-scale bubbling fluidized bed reactor.

provides the schematic representations for the reactor considered in this tutorial. The unit refers to a 250 kW$_{th}$ pilot-scale bubbling fluidized bed reactor located at the Polytechnic Institute of Portalegre, Portugal. The reactor holds a vessel with an internal diameter of 0.50 and 4.15 m height. The bottom bed carries a static height of 0.15 m comprising quartz sand. The employed biomass used for the simulations is Eucalyptus wood (*Eucalyptus globulus*). Dry atmospheric air is used as a gasification agent, being delivered into the reactor at the bottom of the sand bed. The resulting syngas leaves the reactor throughout an outlet placed at the top right corner of the geometry. Simulations were calculated using a time step size of 0.001 s, for a total number of 3000-time steps (3 s).

3. Creating the geometry

By creating the geometry, one is defining the domain and boundary conditions in which the governing equations are to be defined. Here, the reactor's geometry is created from scratch using the ANSYS DesignModeler platform. Before advancing with the design, the user must first simplify the geometry and oust unneeded features. In this tutorial, drawing a rectangle with the same dimensions as the real reactor unit is a possible route to acquire a simplified 2D representation.

3.1 Initiate ANSYS DesignModeler.

3.2 Set the units to Meter.

3.3 Create a new plane by selecting the XYPlane from the Tree Outline and then click on the New Sketch button.

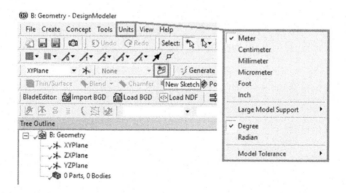

3.4 On the Sketching tab, select Rectangle and draw it by clicking on the origin of the XYPlane.

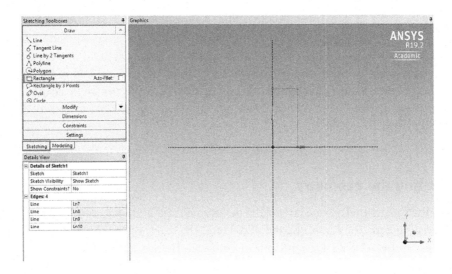

3.5 In the Details View, set the fluidized bed reactor dimensions (0.50 m for length and 4.15 m for height).

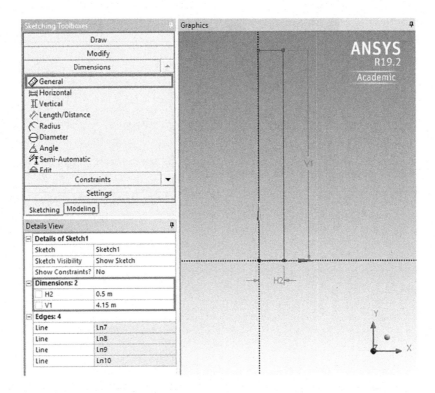

3.6 Click on the Modify toolbox and select Split at Select function to create the outlet boundaries at the top right corner of the Rectangle (0.25 m wide).

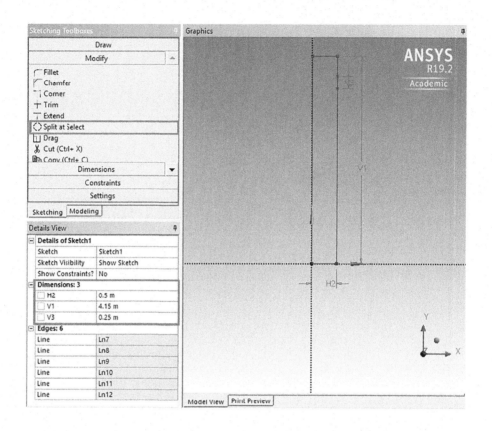

3.7 In order to create the surface of the geometry go to the Concept tab and select Surfaces From Sketches, return to the Modeling tab and select the Sketch icon in the
XYPlane and click Apply in the Details View, click on Generate to create the
surface.

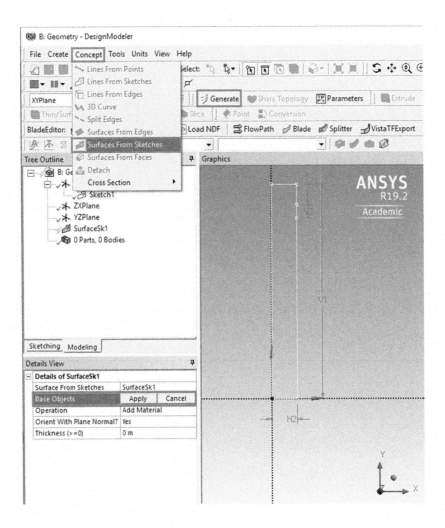

3.8 To name the boundary conditions head to the menu toolbar and on Selection Filter select Edges (for geometry boundaries) or Faces (for geometry surface). Click on the edges and surface of the reactor's geometry and as the boundaries turn green than right click on the boundary and select Named Selection. Specify the air inlet, gas outlet, reactor wall, and the geometry surface as fluid. Having followed all these steps correctly the geometry is now created, the user may now Save the Project.

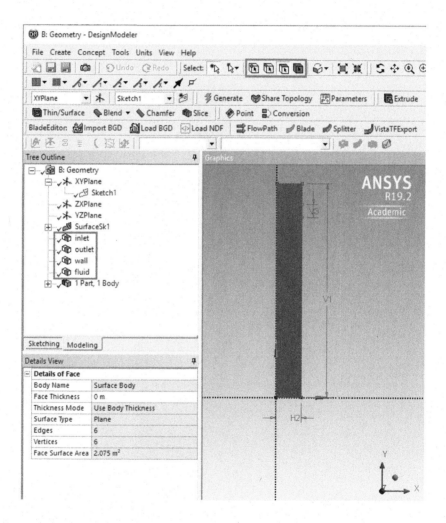

4. Meshing the geometry

The created geometry must now be split into discrete cells (or otherwise known as elements). The mesh influences the accuracy, convergence, and speed of the solution; therefore, the optimal mesh is the one that maximizes accuracy while minimizing the solver run time.

4.1 Initiate ANSYS Meshing and import the created geometry.

4.2 In the Tree, Outline click on Mesh to reveal the Details of "Mesh" and set Physics Preference to CFD and Solver Preference to Fluent.

4.3 In the Sizing toolbox make sure that Capture Curvature and Proximity are enabled. In Quality toolbox set Smoothing to High. Finally, Generate the Mesh.

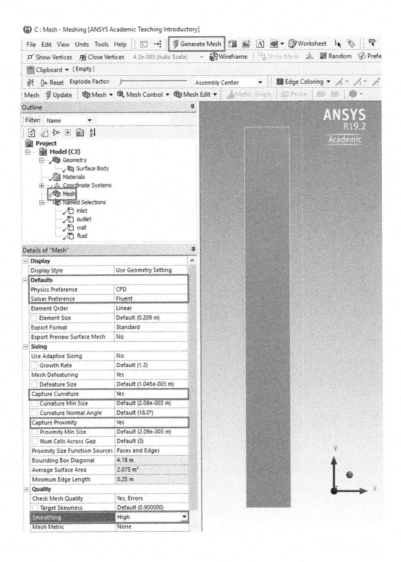

4.4 In the Mesh Control tab select Sizing from the drop list, in the "Sizing" detail options select the face of the rectangle as geometry and press Apply, set 0.0048 m for element size. Return to Mesh Control tab and select Face Meshing and make sure Quadrilaterals is set as default for Method. Lastly, Generate Mesh again.

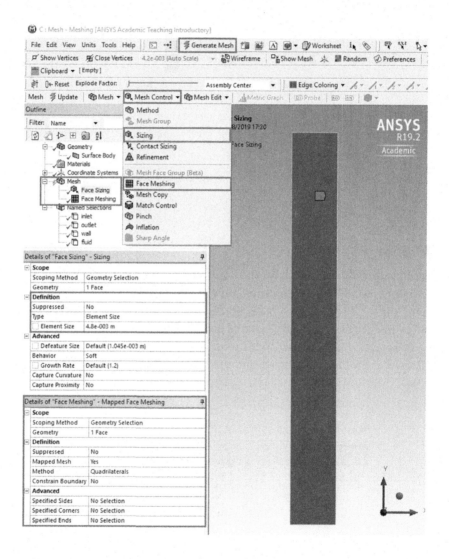

4.5 In order to confirm mesh quality, the user may check the Statistics for the number of Nodes and Elements that contain the mesh. In this tutorial, a mesh containing around 80,000 elements is considered appropriate. Select Element Quality in Mesh Metric from the Quality drop list to check mesh quality. Element quality ranges from 0 to 1, in which higher values indicate higher element quality. Having created a proper mesh quality according to the mesh metrics the user may now Save the Project.

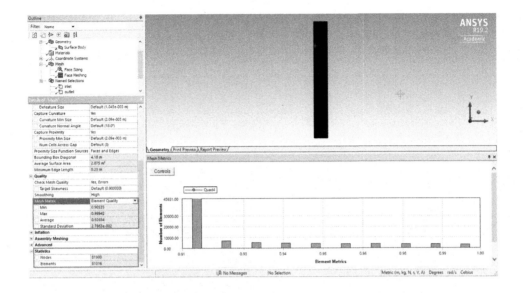

5. Setting-up the solver

Now that the computational mesh for the fluidized bed reactor is created one must configure both physics and solver values based on their knowledge of the fluid application in hands. Here, the user is entitled to set the general solver settings, the definition of physical and chemical mathematical models, material properties, phase creations and interactions, operating cell zone, and boundary conditions.

5.1 Initiate ANSYS Fluent launcher and ensure that the proper options are enabled like so.

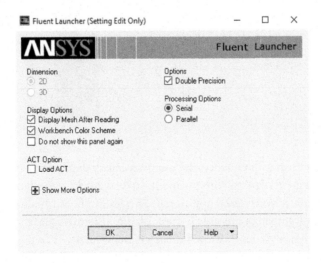

5.2 Select the Setup drop list and start by setting the General options. Click in Check and Report Quality to inspect the Mesh Quality. In Solver settings, enable Type as Pressure-Based, Time as Transient, Velocity Formulation as Absolute, and 2D Space as Planar. Enable the Gravity option in the Y direction and input the gravitational acceleration, -9.81 m/s^2.

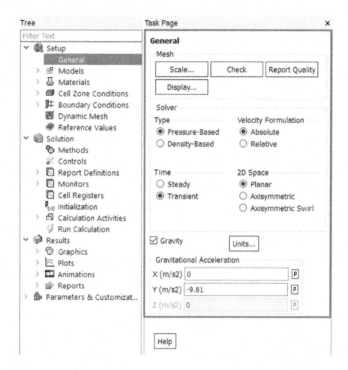

5.3 Define the Models, select Multiphase, enable Eulerian, and set 3 Eulerian Phases (air, bed material, and biomass). Click OK and close the Multiphase Model dialog box. Return to the Models task page, click on Energy and enable Energy Equation. Click OK and close the dialog box. Click on Viscous and enable k-epsilon. Click OK and close the dialog box.

5.4 Compile the user-defined function (UDF), conduct.c, used to define the thermal conductivity for the gas and solid phases. The UDF conduct.c file is made available by ANSYS, Inc. The user can customize this file for better representation in each case. First and foremost, make sure that the conduct.c file and the Fluent case and data files are contained in the same folder. Go to the Parameters and Customization tab, right-click on User Defined Functions and select Compiled. Click the Add… button under Source Files and select conduct.c file and Build. ANSYS Fluent will compile and create a library folder named libudf within the working directory. Details on compiling and interpreting UDF will be broadly covered in another chapter.

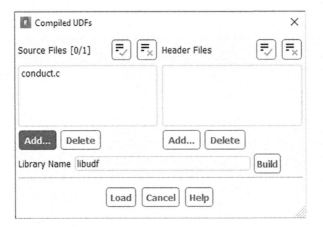

5.5 Go to Materials and modify the properties for the first primary phase, air. Select air then Create/Edit… and set 1.2 kg/m³ for Density, 994 J/kg-K for Cp and user-defined for Thermal Conductivity (select conduct_gas::libudf from the available list), leave all remaining features as default.

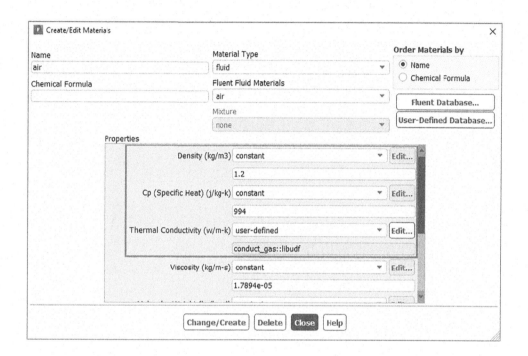

5.6 Return to Materials and modify the properties for solids and biomass. Select solids, here being quartz sand, then Create/Edit… and set 2650 kg/m³ for Density, 830 J/kg-K for Cp and user-defined for Thermal Conductivity (select conduct_solid::libudf), leave all remaining features as default. Create biomass, here eucalyptus wood, and enter 478 kg/m³ for Density, 1374 J/kg-K for Cp, and 0.2881 W/m-K for Thermal Conductivity.

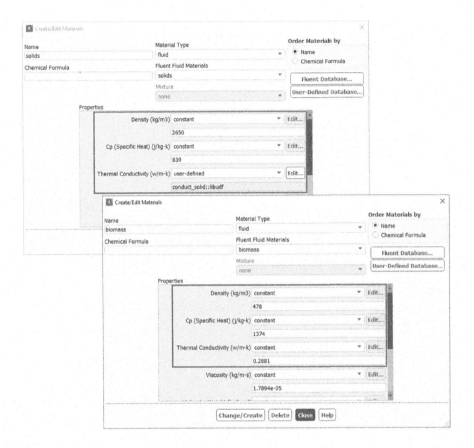

5.7 Define the Phases, set air as primary phase, click on Edit…, enter air for Name and select air from the Phase Material list, click OK, and close the dialog box.

5.8 Define solids (quartz sand) as the secondary phase. Enter solids for Name, select solids from the Phase Material list, enable Granular and Phase Property. Enter 0.0005 m for Diameter, syamlal-obrien for Granular Viscosity, lun-et-al for Granular Bulk Viscosity, algebraic for Granular Temperature, and 0.63 for the Packing Limit. Retain all remaining features as default. Click OK and close.

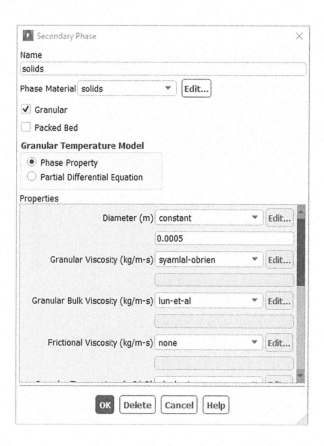

5.9 Define biomass (eucalyptus) also as a secondary phase. Enter biomass for Name, select biomass from the Phase Material list, enable Granular and Phase Property. Enter 0.005 m for Diameter, syamlal-obrien for Granular Viscosity, lun-et-al for Granular Bulk Viscosity, algebraic for Granular Temperature, and 0.63 for the Packing Limit. Retain all remaining features as default. Click OK and close.

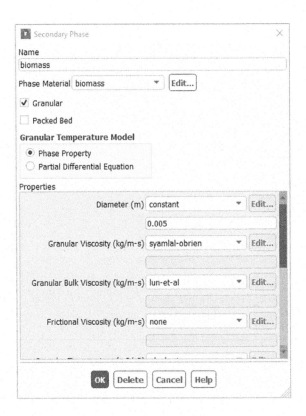

5.10 Now the user must define the phase interactions, click on Interaction... on the Phase dialog box. Set Drag as syamlal-obrien for the solids-air and biomass-air interactions, and syamlal-obrien-symmetric for biomass-solids. Set Heat as gunn for solid-air and biomass-air interactions, and Tomiyama for biomass-solids. Leave all remaining features as default. Click OK and close the dialog box.

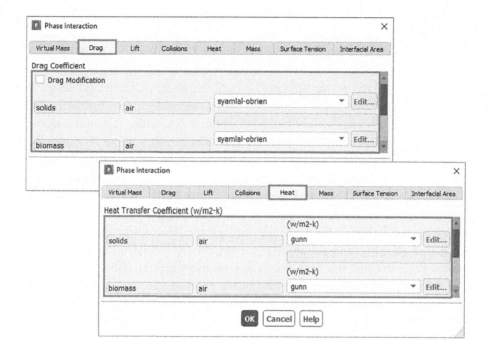

5.11 Regarding the chemical reactions model, go to the Species Model tab, click on Edit right after coal-hv-volatiles-air. The Edit Material tab will be made available. Define a new **biomass-volatiles-air mixture** to deal with the eucalyptus gasification. Manually add the CO and include the char in the fluids box as carbon <solid>. When opening the mixture, the carbon <solid> should fill the solid section. For mathematical purposes, make sure that the nitrogen is the last species under the fluid window. Also, assure that the species are ordered by their predictable amounts.

Select Edit in Reaction and define the Species, Stoichiometric Coefficient, Rate Exponent, and the Arrhenius Rate. In gasification, chemical reactions can be divided into homogeneous and heterogeneous reactions. The first describing the reactions between volatile gases and gasifying agents, while the latter describes the gas species reacting with solid char. Regarding the devolatilization, one must consider that biomass releases moisture during the demoisturization and then it releases volatile matters. In this model, the two Eulerian-Eulerian phases considered are a primary gas phase (air) and a secondary solid phase (quartz sand and eucalyptus biomass). The primary phase contains all gases namely O_2, N_2, H_2O (g), CO, CO_2, H_2, and volatiles, while the secondary solid phases contain char (solid carbon), H_2O (l) and condensed volatiles. Devolatilization occurs after initial drying (along with demoisturization). Here, biomass is thermally decomposed into volatiles, char, and tar [8]. In this tutorial, the biomass volatiles distribution follows a single-rate model approach for moderate and reliable devolatilization rates with little computational effort [6]. In gasification processes, the devolatilization procedure is similar to that of combustion but here one deals with more species. The biomass empirical formula calculation procedure is provided in the previous chapter on combustion.

The homogeneous gas-phase reactions are set by the CO combustion (Eq. 1), water gas shift reaction (Eq. 2), and H_2 (Eq. 3) and CH_4 (Eq. 4) combustion reactions:

$$CO + 0.5O_2 \rightarrow CO_2 \tag{1}$$

$$CO + H_2O \leftrightarrow CO_2 + H_2 \tag{2}$$

$$H_2 + 0.5O_2 \rightarrow H_2O \tag{3}$$

$$CH_4 + 2O_2 \rightarrow CO_2 + 2H_2O \tag{4}$$

The heterogeneous reactions consider the following equations, char combustion (Eq. 5), H_2O (Eq. 6), and CO_2 char gasification (Eq. 7):

$$C + O_2 \rightarrow CO_2 \tag{5}$$

$$CO + H_2O \rightarrow CO + H_2 \tag{6}$$

$$C + CO_2 \rightarrow 2CO \tag{7}$$

The figure below depicts how the user sets the CO oxidation equation. For the sake of simplification, in this chapter the chemical reactions set-up is given in summarized form as the procedure for implementing the chemical model is rather similar to that of combustion, thus, for more detailed and complete information please head to the previous chapter on biomass combustion.

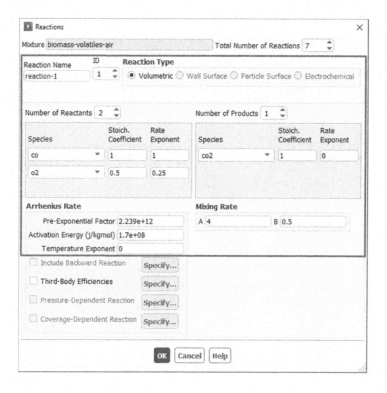

5.12 Ultimately, select Edit in Mechanism and ensure that all chemical reactions are properly selected.

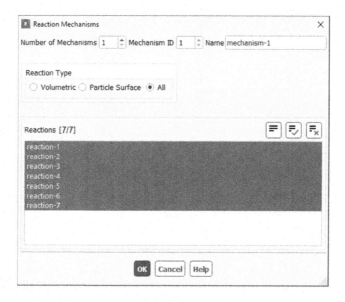

5.13 Set the Boundary conditions, define inlet as velocity–inlet, interior-fluid as interior, outlet as pressure–outlet, and wall as wall. For the Velocity Inlet → Velocity Magnitude as 0.25 m/s → Turbulent Intensity 5% → Hydraulic Diameter 0.26 m (diameter for the distributor plate). Click the Thermal tab and input 500 K. Click Species tab and define 0.23 for O_2. Retain all remaining options as default.

5.14 Regarding the Pressure Outlet, make sure Backflow Direction is Normal to Boundary, define Specification Method as Intensity and Viscosity Ratio, and set 5% for Backflow Turbulent Intensity, and 10 for Backflow Turbulent Viscosity Ratio. Click the Thermal tab and input 300 K. Click Species tab and ensure all species mass fractions are null. Boundary conditions for the wall may be left as default, no actions required.

5.15 Define the Solution Methods → Simple for Scheme → Spatial Discretization, Gradient: Least Squares Cell Based → Pressure: Presto! → Momentum: Second Order Upwind → Turbulent Kinetic Energy: First Order Upwind → Specific Dissipation Rate: First Order Upwind → Species Mass Fractions (vol, O_2, CO_2, H_2O, and CO) set First Order Upwind for all.

5.16 Set the Under-Relaxation Factors → Pressure: 0.3 → Density: 1 → Body Forces: 1 → Momentum: 0.7 → Turbulent Kinetic Energy: 0.6 → Specific Dissipation Rate: 0.8 → Turbulent Viscosity: 0.8 → vol: 0.8 → O_2: 0.8 → CO_2: 0.6 → H_2O: 0.8 → CO: 0.8 → Energy: 0.8.

5.17 Define an adaption register for the bed height of the fluidized bed. In the Adapt tab click Mark/Adapt Cells → Region and set the Region Adaption, X Max 0.50 m for the bed width and Y Max 0.15 m for the bed height, click Mark to refine cells and Close.

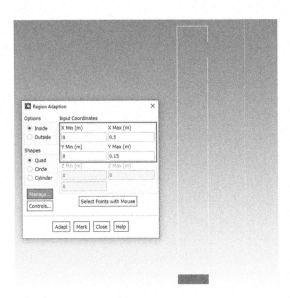

5.18 Initialize the solution by setting 84 for Turbulent Kinetic Energy (m^2/s^2) → 5550 for Specific Dissipation Rate $(1/s)$ → 0.23 for O_2 → 2000 K for Temperature. Press Initialize then Patch the initial volume fraction of solids to the adapted region marked in the fluidized bed reactor. In Patch select Phase solids → Volume Fraction → enters 0.519 for Value → select hexahedron-r0 from Registers to Patch and close.

5.19 Run the calculation by setting the time step size of 0.01 s, for a Number of Time Steps of 3000, to perform a 3 s simulation, enable Extrapolate Variables and Data Sampling for Time Statistics, and click Calculate to initiate the process. All remaining settings may be set as default.

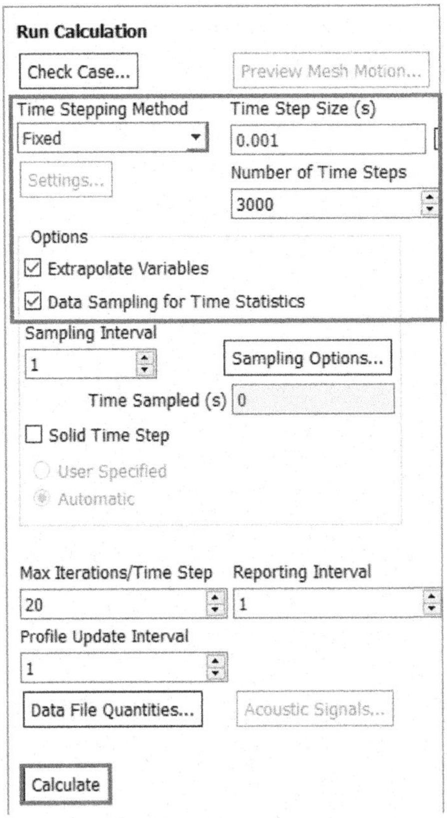

Run Calculation

Check Case...   Preview Mesh Motion...

Time Stepping Method   Time Step Size (s)
Fixed   ▼   0.001

Settings...   Number of Time Steps
                             3000

Options
☑ Extrapolate Variables
☑ Data Sampling for Time Statistics

Sampling Interval
1   Sampling Options...

        Time Sampled (s) 0

☐ Solid Time Step
    ○ User Specified
    ⦿ Automatic

Max Iterations/Time Step   Reporting Interval
20   1

Profile Update Interval
1

Data File Quantities...   Acoustic Signals...

Calculate

6. Postprocessing

The solution once converged the following stage is to process the results. Within the Results task page, a set of most common postprocessing features can be found namely contours, vectors, pathlines, particle tracks, animations, several types of plotting, and reports. Given the numerous features, one may encounter in the ANSYS Fluent post-processing, in this tutorial only a task to visualize the solids volume fraction within the fluidized bed reactor is provided.

6.1 Display the solids volume fraction in the fluidized bed. Click in Graphics → Contours → Set Up. Select solids from the Phase drop-down list, followed by Volume fraction from the Contours of drop-down lists and select the surface the user wishes to analyze, here interior-fluid is chosen. Click Display and close the contours dialog box. This contour display provides the user with information concerning the region inside the reactor in which the change in volume fraction is the greatest.

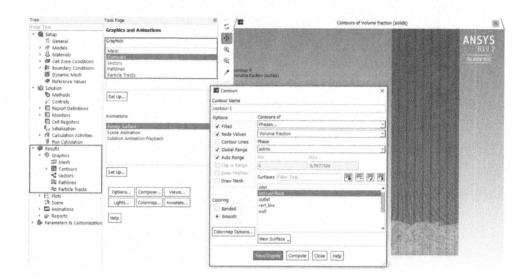

Next, some images illustrating some possible postprocessing contours, one may retrieve from a CFD simulation such as this one:

- Plot contours of coffee husks biomass H_2, CO_2, CH_4, and N_2 molar fractions, respectively [6].

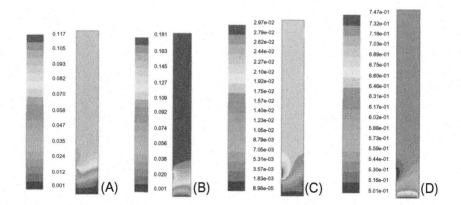

– Plot contours of eucalyptus biomass static temperature during simulation time [2].

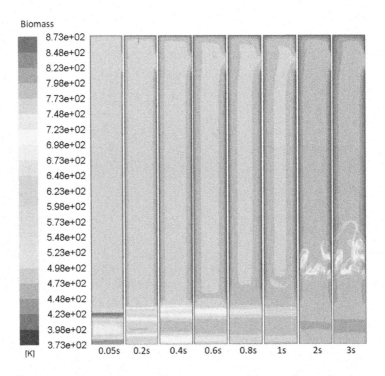

– Plot contours of eucalyptus biomass particle velocity vectors during simulation time [2].

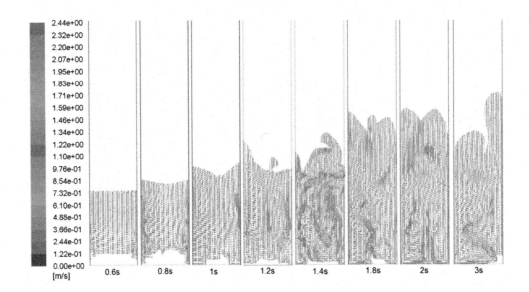

7. Final remarks

A 2D Eulerian-Eulerian approach used to simulate the gasification of eucalyptus biomass in a pilot-scale fluidized bed reactor is provided in this tutorial chapter. Throughout the chapter, the user comprehends to main steps to properly conceive the geometry, define the models, and set the phases interactions and chemical equations mechanism. As seen, ANSYS Fluent provides multiple active options to describe phase interactions, with Drag and Heat settings laying among the most important features to define in fluidized bed systems. The same is true for chemical reactions, as ANSYS Fluent also provided numerous chemical reactions modeling approaches suiting the far most complex process phenomena. Similarly, to the combustion tutorial, the author suggests that the user could converge the system step-by-step (cold gas flow, ignition, multiphase reactions, etc.). In this case, the author used previous data for rapid convergence. Lastly, this tutorial delivers the main steps required for implementing a simple gasification modeling approach while encouraging the user to customize and adapt the solution here resolved to other computational issues that he or she intends to address.

References

[1] J. Cardoso, V. Silva, D. Eusébio, Techno-economic analysis of a biomass gasification power plant dealing with forestry residues blends for electricity production in Portugal, J. Clean. Prod. 212 (2019) 741–753.

[2] J. Cardoso, V. Silva, D. Eusébio, P. Brito, L. Tarelho, Improved numerical approaches to predict hydrodynamics in a pilot-scale bubbling fluidized bed biomass reactor: a numerical study with experimental validation, Energy Convers. Manag. 156 (2018) 53–67.

[3] Q. Zhang, L. Dor, A. Biswas, W. Yang, W. Blasiak, Modeling of steam plasma gasification for municipal solid waste, Fuel Process. Technol. 106 (2013) 546–554.

[4] J. Cardoso, V. Silva, D. Eusébio, P. Brito, R.M. Boloy, L. Tarelho, et al., Comparative 2D and 3D analysis on the hydrodynamics behaviour during biomass gasification in a pilot-scale fluidized bed reactor, Renew. Energy 131 (2019) 713–729.

[5] J. Cardoso, V. Silva, D. Eusébio, Process optimization and robustness analysis of municipal solid waste gasification using air-carbon dioxide mixtures as gasifying agent, Int. J. Energy Res. 43 (2019) 4715–4728.

[6] V. Silva, E. Monteiro, N. Couto, P. Brito, A. Rouboa, Analysis of syngas quality from Portuguese biomasses: an experimental and numerical study, Energy Fuel 28 (2014) 5766–5777.

[7] J. Cardoso, V.B. Silva, D. Eusébio, L. Tarelho, P. Brito, M. Hall, Comparative scaling analysis of two different sized pilot-scale fluidized bed reactors operating with biomass substrates, Energy 151 (2018) 520–535.

[8] J. Khan, T. Wang, Implementation of a demoisturization and devolatilization model in multi-phase simulation of a hybrid entrained-flow and fluidized bed mild gasifier, Int. J. Clean Coal Energy 2 (2013) 35–53.

CHAPTER 5

Gasification optimization using CFD combined with Design of Experiments

1. Introduction

With the environmental and energy players turning away from heavy fossil fuels consumption in the wake of turbulent climate phenomena, the current wave of energy solutions relies mainly on solar, eolic, and biomass sources [1].

While the solar and wind sources continue to impose a strong pace with considerable market share gains on the back of relentless measures oriented to performance improvements, the bioenergy field still relies on obsolete actions that have dubious benefits [2]. There are several key points that could boost an increased bioenergy cost competitiveness:

- Several studies seem to point out that increased actions of scale and standardization in plants allow cost savings up to 20% [3].
- Statistical optimization and variation reduction applied to the technological process and feedstock inhomogeneity could raise the bar in terms of efficiency values (up to 40%). The current average range is about 25% [4].
- Ensuring a secure biofuel supply is perhaps the most important prerequisite for an effective bioenergy power project. This prerequisite depends essentially on both the availability of the feedstock and how effective and stable is the supply chain. In certain cases, depending on the biofuel power plant dimension, on seasonal variation, and on the geographical restrictions, the supply disruption is a likely event, and the need for supplementary purchase is obligatory. Improved supply practices are likely to reduce the fuel cost by 20% [5].

Combustion is a well-known and proven technology but emerging technologies as gasification or pyrolysis might be relevant in short-term and fully performance comparisons are scarce for now. Gasification has been gaining preference over conventional combustion processes due to: (i) the recognition that gaseous fuels have practical advantages over solid fuels, such as handling and application, (ii) the need of renewable fuels that can replace gaseous fossil fuels in distinct applications, and (iii) the flexibility of the gasification processes, due to the various products that can be obtained from the produced fuel gas [6].

Gasification is an extremely complex process to engage given the interplay of several involved phenomena such as hydrodynamics, mass transfer, momentum, energy balance,

Computational Fluid Dynamics Applied to Waste-to-Energy Processes
https://doi.org/10.1016/B978-0-12-817540-8.00005-4

© 2020 Elsevier Inc.
All rights reserved.

and chemical reactions [7]. In addition to the gasifying agent, syngas composition also relies upon many other different interconnected factors namely bed temperature, steam ratio, equivalence ratio, moisture level, feedstock size and chemical composition, among others [8]. Therefore, experimental investigation carries great importance to understand the different entangled phenomena and operational parameters involved in the gasification process while assessing the optimal operating conditions and how these affect the gasifier's performance and syngas composition. Given the intricate nature and the numerous factors participating in the gasification process, employing a conventional one-factor-at-a-time (OFAT) experiment approach requires both strenuous and time-consuming experimentation, neglecting multiple-factor interactions while being unfit to guarantee optimum method conditions [9]. Optimized operation conditions for complex processes such as gasification can be accomplished by employing advanced statistical approaches like Design of Experiments (DoEs) coupled with the Monte Carlo method. DoE allows identifying the most robust combinations of factors by providing a cost-effective and time-efficient experimentation strategy, requiring far less experimental runs than traditional OFAT while simultaneously considering the effects of multiple factors on a given response instead of one at a time [10]. Monte Carlo leverages this optimization process, providing a more reliable and robust estimate by assessing the level of uncertainty [11]. This numerical method, when combined with DoE, allows measuring the variation impact of each factor on the overall performance of the system while investigating the variation of all other involved parameters.

Silva et al. [4] targeted the best operating conditions for several syngas quality indices from a pilot-scale fluidized bed gasifier running Municipal Solid Waste through a full factorial experimental design of 27 runs coupled with a Propagation of Error methodology. The response variation transmitted by the input factors was reduced in order to increase the process CpK and six-sigma standards with a decrease on the tolerance intervals of 20%. The effect of each input factor on the transmitted variation to the response was determined in different scenarios.

This chapter illustrates how these methodologies can be used to find simultaneously the optimal conditions in a gasification process ensuring its robustness.

2. Experimental set-up and substrates characterization

To study the effect of different operating conditions in syngas composition, several tests were performed in a pilot-scale gasification plant. The gasification plant is comprised by an upflow fluidized bed gasifier with a maximum pellet feeding of 75 kg/h. This gasifier is a tubular reactor 0.5 m in diameter and 4.15 m in height, internally coated with ceramic refractory materials and with a bed made of 70 kg of dolomite. Further details concerning the pilot-scale gasification can be found in Ref. [12].

Table 1 Chemical composition used to model the MSW gasification.

Category	% weight	Chemical formula
Cellulosic material	85.42	[a]
Polyethylene	10.99	$(C_2H_4)_n$
Polyethylene terephthalate	2.02	$(C_{10}H_8O)_n$
Polypropylene	0.81	$(C_3H_6)_n$
Polystyrene	0.76	$(C_8H_8)_n$

[a]The considered fraction of cellulose, hemicellulose and lignin was found in Ref. [14].

According to the Portuguese management system [13], during the MSW pretreatment one obtains a refuse-derived fuel (RDF) containing mainly cellulosic materials, paper, wood wastes, and plastic residues. Cellulosic materials are constituted by cellulose, hemi-celluloses, and lignin while plastic residues are mainly composed of polyethylene, poly-styrene, and polyvinyl chloride. All remaining unconsidered products from the MSW treatment seize alternative valorization or elimination routes. Table 1 presents the chem-ical composition for the MSW (as treated by LIPOR) used in ANSYS Fluent to model the gasification process. Contrarily to the plastic residues, whose relative quantities for each monomer come indistinctively listed, the ultimate analysis for the cellulosic materials (cel-lulose, hemicelluloses, and lignin) come undistinguished, their composition was assumed to be similar to the one found in the work of Onel et al. [14].

3. Experimental design

There are several factorial designs being the most known the 2^k and 3^k options and corre-sponding derivations. 2^k designs mean that there are two levels for each k factor and imply an approximately linear response over the range of the factor levels selected. This kind of design is quite usual in the industry and is particularly helpful in a first stage where a large number of factors could be present. Whenever, the developed models suggest a surface with curvature, the two-level designs will no longer give the adequate information and the 3^k designs are the solution. At these circumstances, additional runs are required at new levels of input factors. The simplest approach goes through adding center points to a 2^k design. If one notices sig-nificant variation, then it is wise to add new axial points beyond more center points. Some-times, the range of selected levels is insufficient and the user could assess outside the factorial levels going further out for assessing the curvature. This last approach brings two possible drawbacks: going out from the factorial box could be physically impossible or even unsafe and hit too many levels and consequent too many runs [15].

Any decision should take into account the study specifications, the stage of experi-mentation, the factors restriction, and the possible model linearity or curvature. In gen-eral, typical experiments coming out from the industry lie under quantitative factors such as the temperature or the pressure. This means that these factors are adjusted to any level

Table 2 3^k design for the MSW gasification.

	Gasification runs	ER	CDMR
MSW/air-CO$_2$	1	0.15	0.2
	2	0.25	0.2
	3	0.35	0.2
	4	0.15	0.5
	5	0.25	0.5
	6	0.35	0.5
	7	0.15	0.8
	8	0.25	0.8
	9	0.35	0.8

within the selected range. However, the reader could intend to perform a set of runs using different substrates or even catalysts, and these factors are not quantitative but qualitative or in the DoE jargon, categoric factors. With categoric factors, the user could select only one substrate or another but nothing in the middle. As the reader can guess, the combined use of qualitative and quantitative factors makes the process more complex.

The optimization procedure was engaged considering the results gathered from previous works developed by the author's research team for MSW gasification with air-CO$_2$ mixtures [4, 16]. Overall, nine computer simulations were performed using as input the equivalence ratio (ER) and the CO$_2$-MSW ratio (CDMR). Table 2 presents the full factorial design implemented with all possible selected factors combinations and responses. All remaining operating conditions were kept constant.

3.1 Some considerations

In this particular case, one is coupling the DoE with computational results from CFD simulations. As is easily understandable, CFD simulations always provide the same results and the concept of experimental error does not make sense. Despite that, it is important to state some considerations about the experimental error when the reader is able to perform experimental runs. There is often a clear trend from people who are carrying out the experimental runs to avoid or eliminate the replication of center points. This could turn into great mistakes and the reader is strongly advised to do not do it. Excluding results coming from computational simulations, all the other data should provide for testing of lack of fit (LOF) [17]. LOF presents a relationship between the variation of the replicates and variation of the design points about their predicted values. An F representation for the lack-of-fit test can be as follows:

$$F = \frac{\text{variation between the actual values and the values predicted from the model}}{\text{variation within any replicates}} \quad (1)$$

The lack of such a test compromises the awareness of how the model goes in line with the current response data. To get such information, it is necessary a measure of pure error that comes from replication. To avoid a large number of experimental runs many companies only replicate the center points. At these circumstances, they should get at least four replicates of the center point. Random researches through DoE papers reveal that still a significant number of researchers that do not follow such procedures leading to biased results.

When the user after computing the LOF test detects the model lack-of-fit, this could mean any one of two things: (a) replicates were run as repeated measurements leading to underestimated pure error or (b) replicates have been run correctly and the model is not fitting the design points well.

Sometimes, due to time and laboratory restrictions the center points are gathered in consecutive runs leading to lower errors than expected about the center point. When such occurs, the model fitting must be confirmed applying a battery of additional statistical diagnosing tests. If such tests indicate that the model is robust enough to fit the empirical model under the navigation space, then the lack-of-fit test is no longer a significant test and decisions about the model adequacy should be judged using another statistical analysis.

4. Discussion

4.1 Design of Experiments—Single optimization

The preceding section highlighted that two-level factorial designs do not provide the necessary information to handle the gasification problem. This implies a new level of computational or experimental exploration with additional runs at new levels in the selected parameters.

After selecting the most suitable experimental design, the reader is yet far from generating the RSM plots. First, and foremost, there are some obligatory steps to take in order to ensure that the empirical model is highly predictable within the experimental space. The first step consists of systematically determine how far it is worth going in the polynomial response (full quadratic or cubic polynomials are likely solutions). The sequential model sum of squares (SMSS) is a straightforward method to identify the high-level source of terms responsible for a significant variation in the response. More details about this method can be found in Ref. [18]. Table 3 turns out that the quadratic

Table 3 Sequential model sum of squares (CO_2 generation).

Source	Sum of squares	F value	P-value Prob > F
Quadratic vs 2FI	6.02	350.35	<.0001
Cubic vs Quadratic	0.071	4.81	.0125

source includes all the significant variation for the response. The use of higher sources such as the cubic model is unnecessary, and the response is aliased. Note that there are additional statistical tests to confirm the right source level.

Before to show these statistical measures, it is important to mention that the development of such an empirical model is based on Eulerian-Eulerian simulations under the CFD framework. Any computational-based simulation always provides the same solution for a set of input factors flawing the concept of replicates. As mentioned before, the use of some statistical measures such as lack of fit does not bring any data of interest in such cases, however, measures such as R^2, R^2_{adj}, and R^2_{pred} are still useful. The R^2 measures how well the model is able to fit correctly the experimental data or the computed-based simulations as in the present case. The R^2 value can sometimes be misleading causing overfitting the data. The R^2_{adj} counteracts this overfitting giving a more reliable tool to evaluate the data fitting quality. The R^2_{pred} measures how well the model is able to refit the data when one point is missing. When these measures are close enough a high-quality fit is expected. Table 4 reveals the R measures for the different sources and the results are in line with the SMSS approach. Despite the cubic model presenting high R values, this is a not feasible option as it is aliased. The quadratic model stands out as the most valuable option with considerable high values for all the R measures. Therefore, it is now wise to use a quadratic source for the response.

The higher-order polynomial determined by the sequential model sum of squares (SMSS) suggested that the quadratic model provided the best fitting to the experimental data, including linear (A, B), interaction (AB), and quadratic (A^2, B^2) terms to depict the process variation. A and B terms suited the ER and CMDR factors, respectively, for the air-CO_2 mixtures gasification. A final equation given by the quadratic model to predict the CO_2 generation for the MSW gasification process with air-CO_2 mixtures is given by Eq. (2). Similar analysis and equalization can be carried for all other remaining selected responses.

$$CO_2 = 24.12 + 1.33A + 3.35B + 0.25AB - 0.13A^2 + 0.32B^2 \tag{2}$$

The above equation is in the coded form and it is useful to identify the relative impact of the factors by the signal and magnitude of the coefficients. Positive coefficients mean that increasing the factor leads to a response increase and when their magnitude goes in line with their impact on the response. By default, the high levels of the factors are coded as +1 and the low levels of the factors are coded as −1.

Table 4 Model summary statistics.

Source	R^2	R^2_{adj}	R^2_{pred}
Quadratic	0.9998	0.9998	0.9996
Cubic	1	0.9999	0.9997

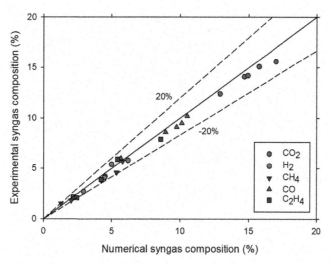

Fig. 1 Experimental versus computational residuals.

The generation of the RSM plots still needs a last step, where the reader should diagnose the residuals or deviations from experimental points. Fig. 1 shows the plot of experimental points for the selected model (CO_2 and other gases) versus the numerical calculation. Note that the plot lines up as expected, and the points hit the diagonal line. The reader has a large list of other diagnostic plots but the previous steps followed in the above sequential order will ensure a great level of confidence in the picked model.

The most straightforward way to evaluate the responses is to examine a contour plot of the fitted model. Fig. 2 shows the contour plot for the single response. When there are only two or three factors under consideration, the interpretation of this kind of plot is easy. However, when the user wants to explore a bunch of variables, the process could become more complex.

It is helpful first to evaluate the single responses without the interaction of the other parameters. This provides relevant information in order to select the best-operating conditions to optimize each one of the responses. Table 5 reveals the operating conditions where this response hits the optimum to consider the applied design.

4.2 Robustness

The following lines proceed with a robust study about the best conditions to reduce the variations transmitted from input variables. This is also a great example to highlight how important is to have a piece of deep knowledge about the process. As mentioned before, RSM provides the optimal operating conditions for different system responses (hydrogen production, CO_2, cold gas efficiency, carbon conversion, among others). This allows generating polynomial functions, which determine the minimum, the maximum or

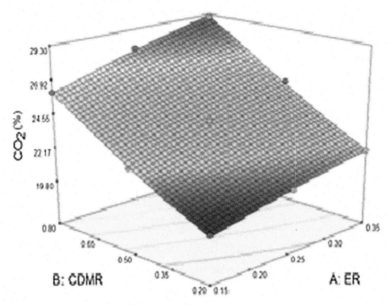

Fig. 2 Contours for the CO_2 response.

Table 5 Maximum optimization prediction values for air gasification.

	Operating conditions for maximized response		
Response	CDMR	ER	Prediction
CO_2 (%)	100	75	29.28

the desired value within a range for each response of interest. However, the previous optimization steps might set the process on a sharp peak of response leading to undesired oscillations and in such manner, the system will not be robust enough for industrial purposes. There are several strategies to find the plateaus on response surfaces such as the Propagation of Error or Monte Carlo, to name a few. These regions are desirable because they are insensitive to variations in factor settings. This kind of analysis allows evaluating the best-operating conditions to target the required syngas properties, as also obtaining a stable syngas generation and a deep understanding of the gasification behavior that is impossible to attain by using traditional experimentation techniques. Furthermore, tolerance intervals can be narrowed, improving the process capability toward six-sigma standards. This is a major step to make the gasification process more repeatable and predictable, leading to significant cost savings.

4.2.1 Some insights about Monte Carlo

In this context, one uses the Monte Carlo methodology, in order to evaluate the response error within the considered space.

The Monte Carlo approach will involve the following steps:

- Determine the standard deviation for each of the input factors. To do so, one must perform some experimental runs. That being impossible, the use of historical data is a typical solution.
- Each one of the input factors will be replaced by probability distribution functions. This means that a range of different values is possible instead of limiting it to just one case. When using historical values, these can be fitted to the best distribution function and the right parameters. Input distributions maybe correlated individually or combined with each other.
- Run the simulation using the Monte Carlo approach meaning that our model is computed thousands of times. The outcomes are recorded and will be used for analysis and for the next step.
- Determine the difference between the response value obtained from the previous point with the one gathered by the empirical RSM model for the same set of input factors.
- One is now able to depict the response error over all the experimental space.
- Proceed with the results analysis. The whole range of possible outcomes is now possible using multiple tools (3D plots, tornado diagrams, histograms, etc.) that will provide uncovered relationships, precedence relations between critical factors and clear highlights about the impact on the outputs.

4.2.2 Robust operating conditions

To accomplish the Monte Carlo analysis, a normal distribution requiring the input of a mean value and standard deviation was considered. This sort of probability distribution is often applied to DoE studies since it produces a realistic behavior of the considered uncertainties [19]. Within this normal distribution, the values were randomly generated under the distribution around the mean value up to their respective standard deviation allowing computing the error created throughout the experimental area. This Monte Carlo simulation process is then repeated for a total of 5000 iterations, approximating the probability distribution of the final result. The employed standard deviations from each of the selected input factors needed to implement the method are given in Table 6. The data source considered was obtained from the experimental historical data

Table 6 Standard deviation for input factors.

Input	Standard deviation
ER	0.01
CDMR	0.01

gathered from the $250\,kW_{th}$ pilot-scale gasification plant as shown in the gasification set–up section.

Fig. 3 presents the 3D plot of the CO_2 generation (at optimal conditions) response error variance for air–CO_2 mixtures gasification as a function of their respective input factors. Identical error variance trends were obtained for the remaining responses, therefore, for simplification purposes, only the CO_2 responses for both processes are provided. This 3D plot combines the optimization procedure with the robust conditions and depicts the most stable operating conditions to achieve the maximum performance of the CO_2 response by providing the operative range at which the response error variance is diminished. The target region to achieve maximum performance is located at the mid-range of their input factors, while their minimum and particularly their maximum ranges return the highest error variances. Major evidence delivered by this kind of information is that the optimal operating conditions required to reach the maximum response earlier determined by single optimization may not always suggest the most stable set of values to operate the system. This is observable from Figs. 2 and 3, as the maximum error variance meets with the optimal operating conditions to obtain a maximized CO_2 generation response, meaning that, in this study, a maximum response condition is not the most stable operating state to work with. This assumption reinforces the need to combine DoE with the Monte Carlo method.

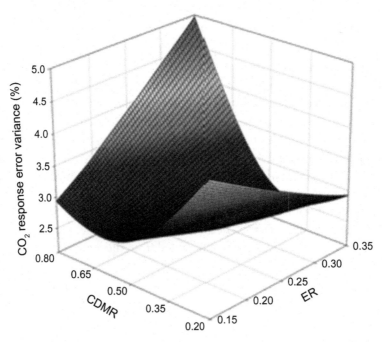

Fig. 3 CO_2 response variance.

Undeniably, this sort of information is particularly valuable in an industrial environment once it helps professionals understanding the level of impact that each factor has on the overall performance of the system, while simultaneously acknowledging the variation of all other involved parameters. In this particular case, combining DoE with the Monte Carlo method grants professionals working in gasification with a set of highly valuable tools and information allowing them to take more secure, reliable and smarter decisions considering the impact on a wide range of factors.

4.2.3 Some final thoughts

This kind of approach and others based in six-sigma philosophy are followed and suggested by leading R&D and industrial entities. Indeed, the evaluation and incorporation of manufacturing best practices, such as 5S, agile manufacturing, six sigma, or statistical process control are necessary not only to the development of high rate manufacturing but also to improve the performance of systems that are already running [20]. In this particular case, the process is improved finding the optimal conditions to maximize an output gas but other responses such as syngas quality indices are also feasible. Furthermore, and beyond an optimization procedure, the engineer can reduce the variability of a process which is nowadays crucial to attend the tight guidelines from the industry.

Advanced approaches like the one presented through this chapter make part of a bunch of strategies to prevent operation instabilities, to find the sweet spots of a complex process, and the ability to follow and sometimes exceed the quality requirements while minimizing quality assurance costs [20].

Beyond the possible economic benefits from using such kind of approaches, it is always important to evaluate how the operating conditions found through these strategies fit the requirements of a real operation. Most of the people usually think that the DoE process only relies on knowledge of statistical methods, but this is not entirely true. Some steps are little related to statistics and are basically focused on the chemical and physical characteristics of the process. The best statistician in the world would not be successful in solving a problem without owing an in-depth knowledge (or at least communicate with who has such knowledge) to choose the factors and levels used in the experiment and in interpreting the results of the statistical analysis. Quite often, the DoE provides a set of appealing solutions that are just not feasible and the user must identify the real sweet spots. This implies a deep knowledge of the process and teamwork between resources with different skills.

ANSYS Fluent includes a statistical package to proceed with statistical analysis. However, the author suggests the use of devoted software to DoE and robust analysis. In general, they present the largest set of features and the user only needs to collect the data from ANSYS simulations. Additionally, the process becomes faster than running in ANSYS. Some examples of these programs are Design Expert, Minitab, or JMP, among others. The reader can also proceed with DoE analysis using the Excel sheets but, in this case, should master the concepts behind the user statistics. Similarly, to DoE, Monte Carlo analysis could be done using Excel but there are available programs such as RISK from Palisade with enhanced features.

5. Conclusions

Economic pressure and the need to target more competitive levels drive organizations to invest in efficient methodologies to get solutions able to provide clear advantages in a very demanding market. In this scenario, statistical approaches emerge as valuable tools to be used in industrial processes. Such approaches range from simple to complex topics such as Design of Experiments, always targeting safer, more repeatable, and profitable solutions. When these methods are combined with CFD models, the user is able not only to predict the process output as optimizing and keeping it under controlled conditions. A 3^k design comprised of nine numerical runs was used to maximize the CO_2 generation as an example.

To ensure that the maximum of a selected response could be obtained without major fluctuations along its working period, one used the Monte Carlo method to find the response variability through the numerical space.

Undeniably, this sort of information is particularly valuable in an industrial environment once it helps professionals understanding the level of impact that each factor has on the overall performance of the system, while simultaneously acknowledging the variation of all other involved parameters. In this particular case, combining DoE with the Monte Carlo method grants professionals working in gasification with a set of highly valuable tools and information allowing them taking more secure, reliable, and smarter decisions.

References

[1] J. Cardoso, V.B. Silva, D. Eusébio, Techno-economic analysis of a biomass gasification power plant dealing with forestry residues blends for electricity production in Portugal, J. Clean. Prod. 212 (2019) 741.
[2] J. Cardoso, V.B. Silva, D. Eusébio, Low carbon economy. An overview, in: V. Silva, M. Hall, I. Azevedo (Eds.), Low Carbon Transition—Technical, Economic and Policy Assessment, Intech-Open, 2018. ISBN: 978-1-78923-970-6. https://doi.org/10.5772/intechopen.70981.
[3] J. Kooroshy, A. Ibbotson, B. Lee, D. Bingham, W. Simons, The Low Carbon Economy GS SUSTAIN Equity Investor's Guide to a Low Carbon World, 2015-25, (2015).
[4] V. Silva, N. Couto, D. Eusébio, A. Rouboa, P. Brito, J. Cardoso, M. Trninic, Multi-stage optimization in a pilot scale gasification plant, Int. J. Hydrog. Energy 42 (2017) 23878.
[5] G.M. Joselin Herbert, A. Unni Krishnan, Quantifying environmental performance of biomass energy, Renew. Sust. Energ. Rev. 59 (2016) 292.
[6] U. Arena, F. Gregorio, Gasification of a solid recovered fuel in a pilot scale fluidized bed reactor, Fuel 117 (2014) 528.
[7] X. Ku, H. Jin, J. Lin, Comparison of gasification performances between raw and torrefied biomasses in an air-blown fluidized-bed gasifier, Chem. Eng. Sci. 168 (2017) 235.

[8] B. Zhang, Z. Ren, S. Shi, S. Yan, F. Fang, Numerical analysis of gasification and emission characteristics of a two-stage entrained flow gasifier, Chem. Eng. Sci. 152 (2016) 227.

[9] H. Yu, H. Yue, P. Halling, Comprehensive experimental design for chemical engineering processes: a two-layer iterative design approach, Chem. Eng. Sci. 189 (2018) 135.

[10] D. Montgomery, Design and Analysis of Experiments, fifth ed., John Wiley & Sons, Inc., USA, 2001.

[11] J. Kleijnen, Design and analysis of Monte Carlo experiments, in: J.E. Gentle, W.K. Härdle, Y. Mori (Eds.), Handbook of Computational Statistics, Springer, 2012.

[12] V.B. Silva, N. Couto, E. Monteiro, P. Brito, A. Rouboa, Analysis of syngas quality from Portuguese biomasses: an experimental and numerical study, Energy Fuel 28 (2014) 5766.

[13] S. Teixeira, E. Monteiro, V.B. Silva, A. Rouboa, Prospective application of municipal solid wastes for energy production in Portugal, Energy Policy 71 (2014) 159.

[14] O. Onel, A.M. Niziolek, M.M.F. Hasan, C.A. Floudas, Municipal solid waste to liquid transportation fuels—part I: mathematical modeling of a municipal solid waste gasifier, Comput. Chem. Eng. 71 (2014) 636.

[15] M. Anderson, P. Whitcomb, RSM Simplified—Optimizing Processes Using Response Surface Methods for Design of Experiments, first ed., Productivity Press, 2005.

[16] N. Couto, V.B. Silva, E. Monteiro, A. Rouboa, Assessment of municipal solid wastes gasification in a semi-industrial gasifier using syngas quality indices, Energy 93 (2015) 864–873.

[17] M. Anderson, P. Whitcomb, Practical Tools for Effective Experimentation, first ed., Productivity Press, 2000.

[18] V.B. Silva, D. Eusébio, J. Cardoso, M. Zhiani, S. Majidi, Targeting optimized and robust operating conditions in a hydrogen-fed Proton Exchange Membrane Fuel Cell, Energy Convers. Manag. 154 (2017) 149.

[19] N. Noguer, D. Candusso, R. Kouta, F. Harel, W. Charon, G. Coquery, A framework for the probabilistic analysis of PEMFC performance based on multi-physical modelling, stochastic method, and design of numerical experiments, Int. J. Hydrog. Energy 42 (2017) 459.

[20] https://energy.gov/sites/prod/files/2014/03/f12/mfg_wkshp_ballard.pdf, last accessed 12 January 2019.

SECTION III

Gasification modeling

CHAPTER 6

Advanced topics—Customization

1. Introduction

Customizing ANSYS Fluent default features can actively enhance the simulation performance. To customize the Fluent solver, users can create and implement their user-defined functions (UDFs). UDFs are C-code-based routines eligible to be loaded in the Fluent solver thus improving the standard features of the code. These make use of Fluent-provided macros and functions to be included during the compilation process.

The implementation of UDF into ANSYS Fluent framework is rather advantageous allowing the user to personalize the setup bringing the solution closer to its particular modeling needs. ANSYS Fluent concedes the user to tailor plentiful of its standard features by UDF inclusion, namely boundary conditions, material properties, surface and volume reaction rates, source terms in transport equations, diffusivity functions, etc. Despite its wide range of applications, UDF inclusion conveys some limitations as not all solution variables or Fluent models can be tailored by UDF, which would demand additional solver capabilities.

The proper prediction of fluidization behavior is a long-standing issue in fluidized bed industrial processes. ANSYS Fluent framework provides a set of rigorous mathematical modeling tools for describing the complex interphase momentum in multiphase phenomena such as combustion and gasification. Still, the intrinsic drag laws employed to describe these phenomena remain semiempirical, lacking to provide accurate behavior. Therefore, it is of utmost importance to customize the drag law to correctly predict the fluidization conditions.

To quantify this, an identical approach was applied to improve a fluidized bed reactor hydrodynamics acting over the drag model. Its application revealed a significant effect upon the model validation by providing a better agreement between the numerical and experimental results gathered from a pilot-scale fluidized bed reactor, as seen in the fluidization curves depicted in Fig. 1. Here, the drag model adjustments were achieved by tuning the default drag coefficients parameters within their governing equations permitting to alter the general correlations in the Fluent database so to fit the intended modeling requirements and to better agree with the experimental setup [1].

Overall, UDF implementation aids users and researchers by minimizing errors and deviations, assuring a better agreement between experimental and numerical results, strengthening the mathematical model accuracy and predictability.

Computational Fluid Dynamics Applied to Waste-to-Energy Processes
https://doi.org/10.1016/B978-0-12-817540-8.00006-6

© 2020 Elsevier Inc.
All rights reserved.

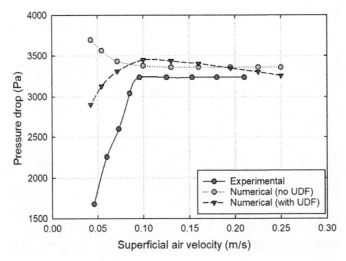

Fig. 1 Comparison between experimental and numerical fluidization curves with and without UDF implementation.

In this sense, the purpose of this section is to provide the user with the required set of skills to successfully implement and run customization functions in the ANSYS Fluent solver.

2. Sample function

Providing a glance at a UDF layout, ANSYS Fluent Inc. hands out some sample problems containing already created functions. To create UDF from scratch, the user must possess some programming experience with C. Without delving too deeply into the function operation details, the following UDF was built to introduce a parabolic profile at the inlet in a turbine vane. Additional information concerning this, and others, UDF can be found elsewhere [2]. The C source code is shown below:

```
/********************************************************************
vprofile.c
UDF for specifying steady-state velocity profile boundary condition
********************************************************************/
#include "udf.h"
DEFINE_PROFILE(inlet_x_velocity, thread, position)
{
real x[ND_ND]; /* this will hold the position vector */
real y;
```

```
face_t f;
begin_f_loop(f, thread)
{
F_CENTROID(x,f,thread);
y = x[1];
F_PROFILE(f, thread, position) = 20. - y*y/(.0745*.0745)*20.;
}
end_f_loop(f, thread)
}
```

3. Interpreting and compiling UDF

UDF once created can be executed within the Fluent framework either by being interpreted or compiled. Choosing to either interpret or compile UDF demands acknowledging some major differences. Interpreted UDF runs slower and do not require a C compiler, while compiled UDF runs faster than interpreted UDF yet these oblige an installed C compiler. Overall, interpreted UDF is preferred for smaller and straightforward functions, while compiled UDF is suitable for more complex functions compelling significant CPU requirements. The steps for interpreting or compiling UDF are detailed throughout this section.

3.1 Interpreting UDF

3.2 Start Fluent. Make sure the UDF file is in the same working directory as the case and data files.

3.3 Read (or set up) the case file.

3.4 Open the Interpreted UDFs panel by selecting Parameters & Customization > User Defined Functions (right-click) > Interpreted...

3.5 In the Interpreted UDFs panel, browse the source file to interpret (e.g., "bp_drag.c"), specify the C preprocessor CPP Command File Name as CPP, keep the default 10000 for Stack Size, and click Interpret. One may close the panel as soon as the interpretation is over. Mind the Fluent solver Console for any reported errors.

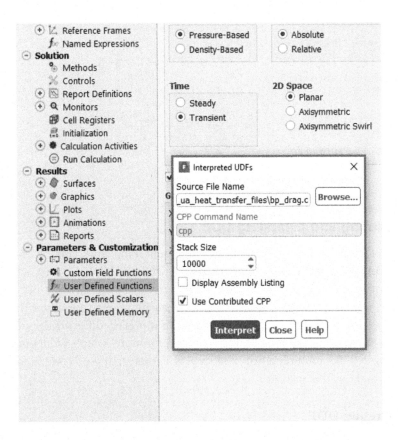

3.6 The source file once interpreted the user must now set up the UDF. For instance, if the interpreted UDF is set to customize the default drag coefficients, one must assign the Drag Coefficients as user-defined in the Phase Interaction panel. This procedure allows calling the UDF to be loaded during the simulation.

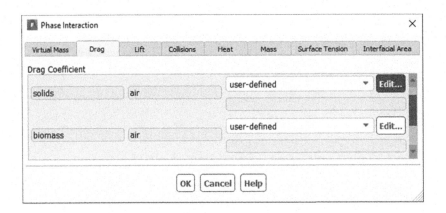

3.7   Finally, writing the case file will save the interpreted function and automatically interpret it when the case file is subsequently read.

3.8   Compiling UDF

3.9   Start Fluent. Make sure the UDF file is in the same working directory as the case and data files.

3.10   Read (or set up) the case file.

3.11   Open the Compiled UDFs panel by selecting Parameters & Customization > User Defined Functions (right-click) > Compiled…

3.12   In the Compiled UDFs panel, click in Add… under the Source Files tab and browse the source file to compile (e.g., "conduct.c").

3.13   Set the Library Name or leave it as default "libudf." Make sure that the UDF file is in the working directory containing the case and data files. Click on Build. This task builds a shared library for the source file (UDF).

3.14   Once the building is successful, click on Load. This task will load the created shared library. Mind the Console for any reported errors.

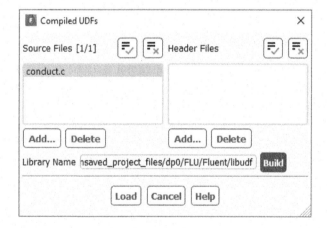

3.15   Write the case file to save the compiled function and automatically load the library the next time you read the case file.

3.16   If the compiled UDF is set to customize the gas-solid thermal conductivity parameters within their governing equations, one must assign the Thermal Conductivity as user-defined in the Materials panel to hook the UDF to the Fluent solver during the simulation process.

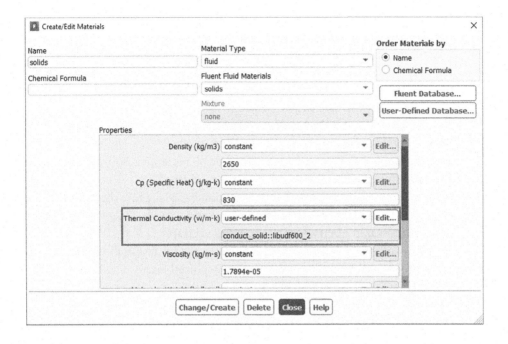

4. Creating a custom field functions

In addition to broad customization employing UDF files, the user may also set some basic field variables and define his field functions to tailor contour and vector display, XY plots, etc. To do so, access the Custom Field Function Calculator toolbox in Parameters & Customization. Here, one can use default field variables, previously defined calculator functions and calculator operators to create new functions, all by using simple calculator operators. Creating custom field functions come particularly handy every time the default Fluent variables are insufficient to describe the issue in hand.

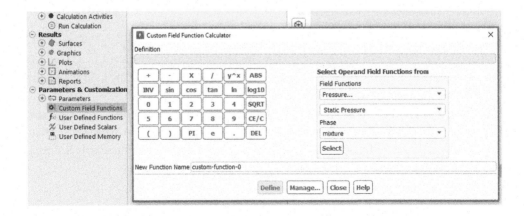

References

[1] J. Cardoso, V. Silva, D. Eusébio, P. Brito, L. Tarelho. Improved numerical approaches to predict hydrodynamics in a pilot-scale bubbling fluidized bed biomass reactor: a numerical study with experimental validation, Energy Convers. Manag. 156 (2018) 53–67.

[2] Fluent Inc., Fluent 6.1 UDF Manual, (2003).

CHAPTER 7

Advanced topics—Postprocessing

1. Introduction

While performing a CFD analysis, the user attains quantitative and qualitative data regarding the fluid flow performance of the analyzed system. Following the results prediction, postprocessing tasks are required, handing over a complete insight into the user's simulation results.

ANSYS tools dispose of two possible routes to postprocess CFD results, one is through Fluent postprocessing tools, integrated into the Fluent solver, while the other is through CFD-Post application which can be run as a stand-alone postprocessor or within ANSYS Workbench. Both routes convey many tools for analyzing CFD results such as multiple graphic options from isosurfaces, vector plots, contour plots, to streamlines and path lines, also as various plotting features, reports, and animations. Furthermore, one may also postprocess any desired user-defined quantity by implementing user-defined functions (UDFs) or through custom field functions. Usually, most of these tools operate on geometry lines or surfaces created by the user within the Fluent framework [1].

Comparatively, Fluent postprocessing tools allow the user to promptly review and manage the simulated data within the solver, standing as a more convenient path to set the most basic postprocessing operations. On the other hand, CFD-Post application allows insightful high-end graphics bestowing far more powerful and sophisticated postprocessing capabilities, including three-dimensional (3D) viewer files, user variables, automatic HyperText Markup Language (HTML) report generation, and case comparison, providing researchers with a handful of valuable high-quality visuals for disseminating and comprehending the most complex phenomena.

Therefore, and as postprocessing skills are best learned in a hands-on manner, this tutorial chapter focuses on introducing the capabilities of CFD-Post application by providing the reader with the required tools to successfully grasp the features available within the CFD-Post scheme in treating and visualizing the simulation results from biomass gasification processes in a pilot-scale bubbling fluidized bed reactor.

2. Problem description

This tutorial is based on a computational two-dimensional (2D) Eulerian-Eulerian approach set to simulate the hydrodynamics of a biomass gasification process in a

Computational Fluid Dynamics Applied to Waste-to-Energy Processes
https://doi.org/10.1016/B978-0-12-817540-8.00007-8
© 2020 Elsevier Inc.
All rights reserved.

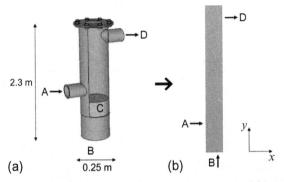

Fig. 1 Pilot-scale fluidized bed reactor domain: (a) Reactor schematics and (b) simplified 2D geometry.

pilot-scale bubbling fluidized bed reactor. The unit refers to a 75 kW_{th} pilot-scale bubbling fluidized bed reactor located at the University of Aveiro, Portugal (Fig. 1a). The reaction chamber consists of an internal diameter of 0.25 and 2.3 m height. The bottom bed carries a static height of 0.23 m comprising 17 kg of quartz sand with particles ranging between 355 and 710 μm. Eucalyptus wood (*Eucalyptus globulus*) is employed as biomass during the simulation. Dry atmospheric air is used, being fed into the reactor throughout the bottom of the sand bed. The produced gas (syngas) leaves the reactor throughout an outlet placed at the top right corner of the geometry. The computational geometry domain refers to a simplified 2D representation designed with the same dimensions than the pilot-scale unit, as shown in Fig. 1b. Simulations were calculated using a time step size of 0.001 s, for a total number of 1400 time steps (1.4 s). The number of time steps was considered enough to perceive the main features of the hydrodynamic behavior.

3. Solution display in CFD-Post application

3.1 Initiate CFD-Post application either within the CFX-Solver Manager or within ANSYS Workbench.

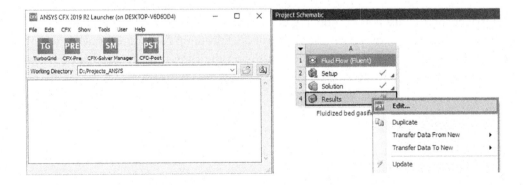

3.2 If the user launched the CFD-Post within the ANSYS Workbench, all solution results become automatically available for postprocessing analysis and no additional actions are required. Yet, when running CFD-Post application as stand-alone within the CFX-Solver Manager, the user must Load the Results File. Within the Fluent solver, the user can create CFD-Post compatible files, which can be imported into CFD-Post application for postprocessing the solution results. On the Solution tab select Calculation Activities and beneath the Automatic Export box click on Create and select Solution Data Export.

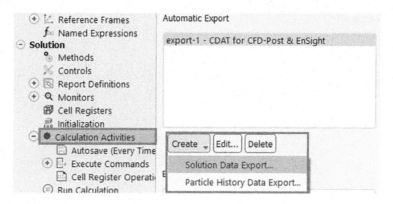

3.3 An Automatic Export window is now available. Here, select the File Type as CFD-Post Compatible or CDAT for CFD-Post & EnSight (for ANSYS R1 onwards version) from the drop-down list, and highlight the cell zones and variables (Quantities) one desires to export for further analysis. Lastly, set Export Data Every 50 time steps and define the file name. Leave all remaining options as default. This procedure will create a CFD-Post compatible file with the extension .cdat suitable to import into CFD-Post application.

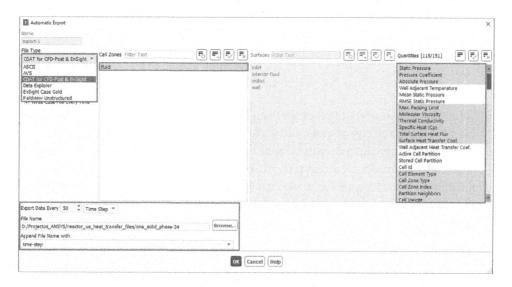

3.4 Having imported the solution results into the CFD–Post application take a moment to familiarize with controls disposition. In short, on the left boxes, CFD-Post displays the model objects and details, at the center the geometry alongside with multiple viewer options, and on top all sorts of tabs related to other graphical options. Start by clicking on the Z-axis on the viewer triad to get a front view of the geometry.

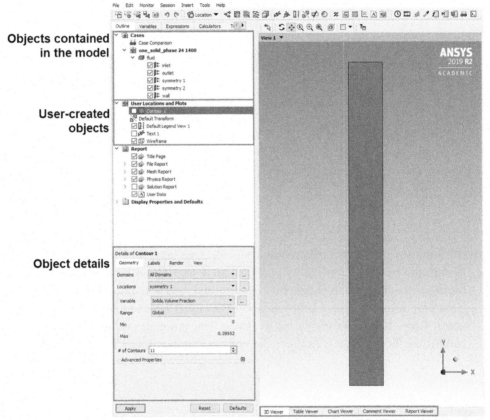

Multiple viewers (3D, Table, Chart, etc.)

3.5 While studying fluidized bed gasification hydrodynamics, contouring solids volume fraction, solids velocity vectors, and static pressure come as rather useful variables to assess. Hereupon, the user will be led on how to set the CFD-Post application to enable these variables. In this tutorial, only graphical results will be displayed as the plotting and charting features contained within the Fluent postprocessing solver are sufficient to obtain good plotting representations, unlike its general graphical abilities.

3.6 To create a contour, start by clicking on the Contour button, name it Solids Volume Fraction and press OK. In the Geometry tab, within the Details toolbox, set symmetry 2 for Locations, then set Solids.Volume Fraction for Variable and press Apply. All remaining features may be left as default. Depicting solids volume fraction comes in handy in reactor hydrodynamics analysis as it allows to visualize the solids disposition along the bed height and assess over the hydrodynamic behavior of the solids throughout the gasification process.

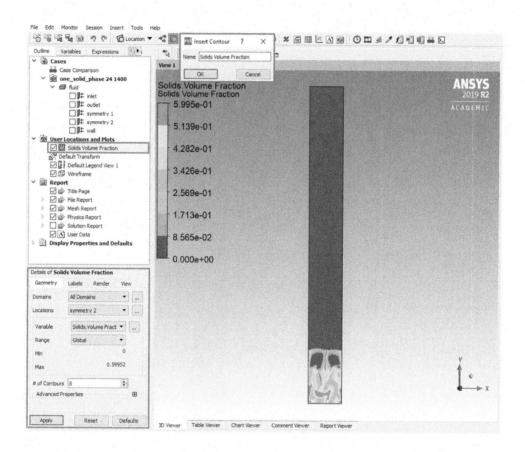

3.7 Following the same outline, to create a static pressure contour, click on the Contour button, name it Static Pressure and press OK. In Details > Geometry tab set symmetry 2 for Locations, then set Pressure for Variable and press Apply. All remaining features may be left as default. In order to display the Pressure contour, the user must disable the Solids Volume Fraction and enable Static Pressure contour in the User Locations and Plots drop-down list. Monitoring the numerical static pressure distribution throughout the gasification process is an important step as it allows to measure the pressure drop of the system and further compare it with experimental measurements.

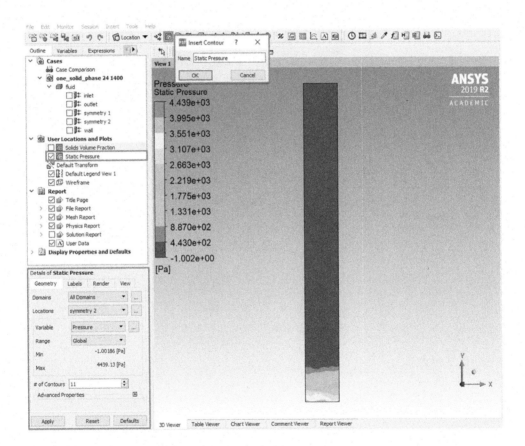

3.8 To create the solids velocity vectors, click on the Vector button, name it Solids Velocity and press OK. In Details > Geometry tab set symmetry 2 for Locations, Vertex for Sampling, Reduction Factor for Reduction, increase the default Factor value from 1 to 2 for better visualization of the vectors, Solids.Velocity for Variable and press Apply. All remaining features may be left as default. Solid particles movement within the fluidized bed is promoted by the airflow in the bed height. Solids velocity vectors allow measuring the solid particles velocity along the bed height and determine the locations in which these reach increased velocity either being at the center or near-wall regions.

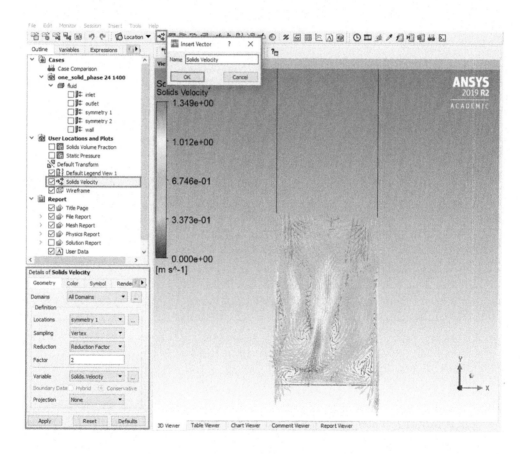

3.9 From the File toolbar, select Save As to the save the postprocessing progress into a directory of desire in a .cst file. To save the created contours, select File > Save Picture, then select the File name and directory. If desired, the user may set the background to white for better visualization and further inclusion in article or report by enabling White Background. Leave all remaining options as default.

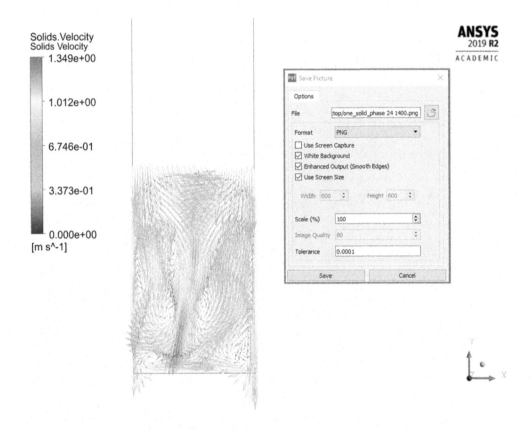

3.10 A comparison between the solids volume fraction and solids velocity vectors captured from the Fluent postprocessing solver and the same from the CFD-Post application shows the enhanced graphical capabilities of the latter.

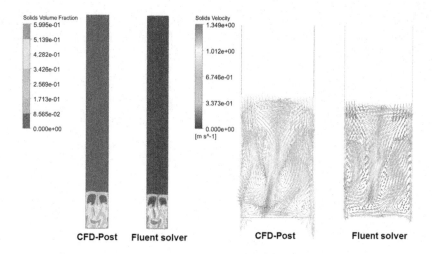

CFD-Post   Fluent solver   CFD-Post   Fluent solver

4. Creating an animation video

The CFD-Post application allows creating animation videos allowing to fully grasp the whole simulation process in a time-fluid manner. Moreover, animations are the ultimate aid to fully disseminate one's work capturing far more easily the interest of the viewers in a rather entertainable and interactive fashion.

4.1 First select the variable of interest to describe in the video, here solids volume fraction is enabled.

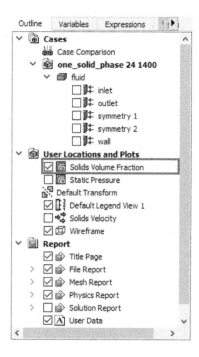

4.2 Click on the Animation icon to open the Animation dialog window. Select Time-
step Animation, enable Save Movie, chose the desired file name and directory, and
set MPEG4 for Format.

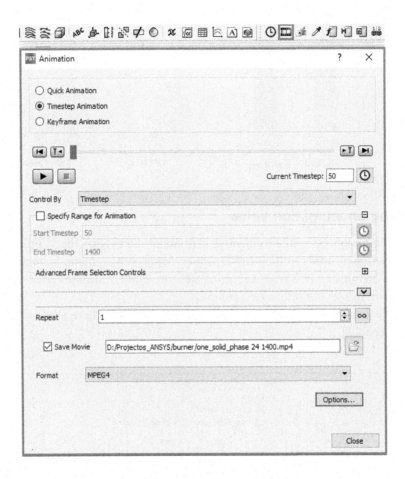

4.3 Click on Options to open the Animation Options window, improve the video quality by setting HD Video 1080p in Image Size. In the Advanced tab, make sure Quality is set to Highest.

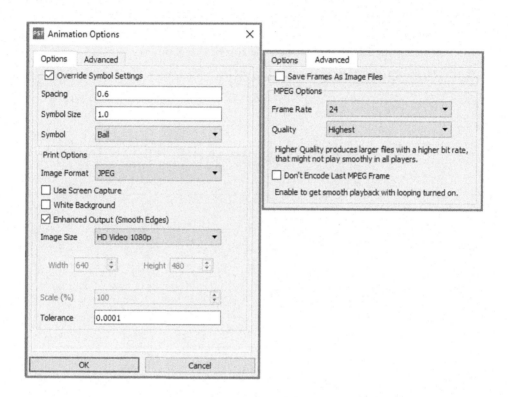

4.4 Before initiating the animation video, a useful feature is to display the simulation time during the animation video. To do so, click on the Text icon, name it Simulation Time and press OK. In the Details > Definition, insert "Time =" in the Text String, enable Embed Auto Annotation and in Type set Time Value from the drop-down list and press Apply. A text string will now be made available on the top of the reactor. Finally, with the viewing cursor, adjust the geometry in case the text string appears slightly above the top of the geometry.

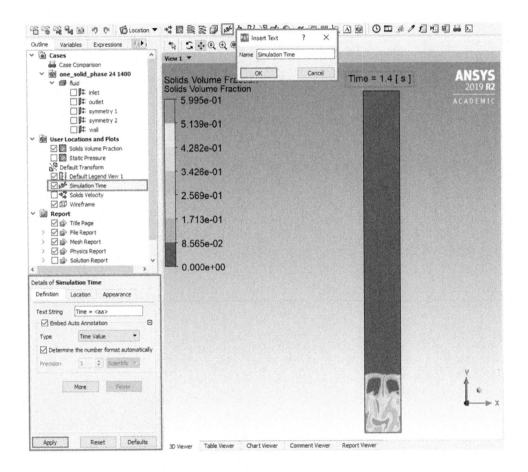

4.5 Having performed all these tasks, the user may now press the play button in the Animation window to start recording the animation video. The video is automatically saved in the chosen directory as .mpeg4 file. Use a video play software of choice to delay and/or repeat the animation video as the user sees fit.

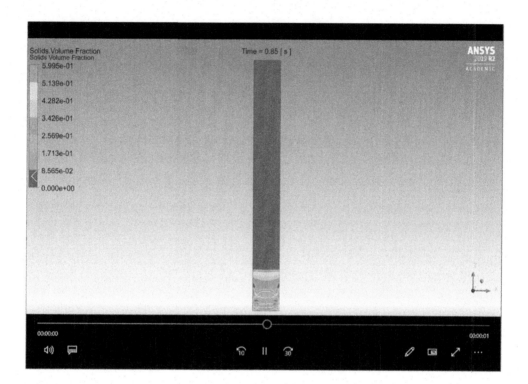

5. Final remarks

Concerning fluidized bed gasification, the use of CFD-Post application allows producing visual data with higher quality, assisting to better visualize and understand the complex flow phenomena within the reactor. Indeed, the ANSYS CFD-Post features are immense and are best learned in a hands-on manner, the experience will lead users to take their results visualization and analysis to the next level by taking the maximum potential of such a broad application.

Reference

[1] ANSYS, ANSYS CFD-Post User's Guide, Release 150, ANSYS, Inc., 2013.

Appendix

Appendix A

A.1 Mathematical model

To simulate the biomass gasification process, a multiphase (gas and solid phase) model from a Fluent database was used. In the comprehensive two-dimensional (2D) numerical model, the gas phase is treated as continuous, and the solid phase is described through an Eulerian granular model. Interactions between both phases were modeled as well, since both phases exchange heat by convection, momentum (given the drag between gas phase and solid phase), and mass (given the heterogeneous chemical reactions).

A1.1 Mass balance model

$$\frac{\partial}{\partial t}\left(\alpha_g \rho_g\right) + \nabla \cdot \left(\alpha_g \rho_g \vec{v}_g\right) = S_{gs} \tag{1}$$

The continuity equation for the gas and solid phases is given by

$$\frac{\partial}{\partial t}\left(\alpha_s \rho_s\right) + \nabla \cdot \left(\alpha_s \rho_s \vec{v}_s\right) = S_{sg} \tag{2}$$

The biomass feed reacts with the oxygen steam and carbon dioxide to change solid phase into gas phase, so S which is defined as the mass source term due to heterogeneous reactions, can be expressed as follows:

$$S_{sg} = -S_{gs} = M_c \sum \gamma_c R_c \tag{3}$$

The ideal gas behavior was considered for computing the gas-phase density:

$$\frac{1}{\rho_g} = \frac{RT}{p} \sum_{i=1}^{n} \frac{Y_i}{M_i} \tag{4}$$

The solid-phase density was assumed as constant.

A1.2 Momentum equations

The momentum equation of the gas phase is

$$\frac{\partial}{\partial t}\left(\alpha_g \rho_g \vec{v}_g\right) + \nabla \cdot \left(\alpha_g \rho_g \vec{v}_g \vec{v}_g\right) = -\alpha_g \cdot \nabla p_g + \nabla \cdot \alpha_g \bar{\tau}_g + \alpha_g \rho_g \vec{g} + \beta\left(\vec{v}_g - \vec{v}_s\right) + S_{gs} U_s \tag{5}$$

The momentum equations of solid phase can be written as

$$\frac{\partial}{\partial t}\left(\alpha_s\rho_s\vec{v}_s\right) + \nabla\cdot\left(\alpha_s\rho_s\vec{v}_s\vec{v}_s\right) = -\alpha_s\cdot\nabla p_s + \nabla\cdot\alpha_s\bar{\tau}_s + \alpha_s\rho_s\vec{g} + \beta\left(\vec{v}_g - \vec{v}_s\right) + S_{sg}U_s \quad (6)$$

A1.3 Turbulence model

The standard k-ε model in ANSYS Fluent has become the workhorse of practical engineering flow calculations in the time since it was proposed by Launder and Spalding [1]. It is a semiempirical model, and the derivation of the model equations relies on phenomenological considerations and empiricism. The selection of this turbulence model is appropriate when the turbulence transfer between phases plays a predominant role as in the case of gasification in fluidized beds.

The turbulence kinetic energy k and its rate of dissipation ε are obtained from the following transport equations:

$$\frac{\partial}{\partial t}(\rho k) + \frac{\partial}{\partial x_i}(\rho k_{u_i}) = \frac{\partial}{\partial x_j}\left[\left(\mu + \frac{\mu_t}{\sigma_k}\right)\right] + G_k + G_b - \rho\varepsilon - Y_M + S_k \quad (7)$$

$$\frac{\partial}{\partial t}(\rho\varepsilon) + \frac{\partial}{\partial x_i}(\rho\varepsilon_{u_i}) = \frac{\partial}{\partial x_j}\left[\left(\mu + \frac{\mu_t}{\sigma_\varepsilon}\right)\frac{\partial\varepsilon}{\partial x_j}\right] + C_{1\varepsilon}\frac{\varepsilon}{k}(G_k + C_{3\varepsilon}G_b) - C_{2\varepsilon}\rho\frac{\varepsilon^2}{k} + S_\varepsilon \quad (8)$$

A1.4 Granular Eulerian model

In the granular Eulerian model, stresses in the granular solid phase are obtained by the analogy between the random particle motion and the thermal motion of molecules within a gas accounting for the inelasticity of solid particles. As in a gas, the intensity of velocity fluctuation determines the stresses, viscosity, and pressure of the granular phase. The kinetic energy associated with velocity fluctuations is described by a pseudothermal temperature or granular temperature, which is proportional to the norm of particle velocity fluctuations.

The conservation equation for the granular temperature, obtained from the kinetic theory of gases, takes the following form:

$$\frac{3}{2}\left[\left(\frac{\partial(\rho_s\alpha_s\Theta_s)}{\partial t} + \nabla\cdot\left(\rho_s\alpha_s\vec{v}_s\Theta_s\right)\right)\right] = (-P_s\bar{I} + \bar{\tau}_s):\nabla\left(\vec{v}_s\right) + \nabla(k_{\Theta_s}\nabla\Theta_s) - \gamma_{\Theta_s} + \varphi_{gs} \quad (9)$$

The diffusion coefficient for granular energy was computed by using the following equation due to Syamlal et al. [2]:

$$k\theta_s = 15d_s/4(41 - 33\omega)\rho_s\alpha_s\sqrt{\theta_s\pi}\left[1 + \frac{12}{5}\omega^2(4\omega - 3)\alpha_sg_{0,ss} + \frac{16}{15\pi}(41 - 33\omega)\omega\alpha_sg_{0,ss}\right]$$

$$(10)$$

where $\omega = 1/2(1 + e_{ss})$; d is the biomass particle diameter and s is the solid phase.

The granular energy dissipation can be computed by using the expression derived by Lun et al. [3]. When the granular flow has a smaller volume fraction than the maximum possible value, a solid pressure is considered for the pressure gradient term in the momentum equation, which includes a kinetic term and a particle collision term:

$$p_{s} = \alpha_{s}\rho_{s}\theta_{s} + 2\rho_{s}(1 + e_{ss})\alpha_{s}^{2}g_{0,ss}\theta_{s} \tag{11}$$

The radial distribution function allows the different levels of compressibility. It works as a correction factor that gives the probability of collisions when the granular phase goes to denser states. ANSYS Fluent provides empirical relations for the radial distribution function, when there is one solid phase.

A1.5 Granular Eulerian model
A1.5.1 Devolatilization
Devolatilization is a process where moisture and volatile matters are driven out from biomass by heat. No devolatilization process is included in ANSYS Fluent considering the Eulerian-Eulerian method. To develop a reliable gasification model, it is necessary to include a devolatilization model in the Fluent code.

Biomass is thermally decomposed into volatiles, char, and ash, in agreement with the following equation:

$$biomass \longrightarrow char + volatiles + steam + ash \tag{12}$$

Note that ash appears for illustration only; the simulation does not consider ash content in the solid phase. In the present work, the volatile matter is composed of the following species:

$$volatiles \longrightarrow \alpha_1 CO + \alpha_2 CO_2 + \alpha_3 CH_4 + \alpha_4 H_2 \tag{13}$$

The biomass mixture composition is determined based on the proximate and elemental analysis.

Since there are no data on the exact distribution of the volatiles in biomass, the single-rate model developed by Badzioch and Hawsley [4] was adopted. The single-rate model produced moderate and reliable devolatilization rates with little computational effort. In addition, a demoisturization equation was also considered. These two pseudo-heterogeneous reactions were modeled with a single reaction rate in agreement with the Arrhenius law.

A1.5.2 Homogeneous gas-phase reactions
The modeling of the homogeneous gas-phase reactions should consider both the kinetic and the turbulent mixing rate effects [5]. The gas phase is conditioned by the effect caused

by the chaotic fluctuations of the solid particles. This turbulent flow leads to velocity and pressure fluctuations on the gaseous species. Fluent provides the finite-rate/eddy-dissipation model, which considers both the Arrhenius and the eddy-dissipation reaction rates. The water-gas shift reaction and the CO, H_2, and CH_4 combustion reactions were considered, respectively:

$$CO + H_2O \leftrightarrow CO_2 + H_2 \tag{14}$$

$$CO + 0.5O_2 \rightarrow CO_2 \tag{15}$$

$$H_2 + 0.5O_2 \rightarrow H_2O \tag{16}$$

$$CH_4 + 2O_2 \rightarrow CO_2 + 2H_2O \tag{17}$$

The Arrhenius rates for each one of these reactions can be expressed as follows, respectively:

$$r_{CO\ combustion} = 1.0 \times 10^{15} \exp\left(\frac{-16,000}{T}\right) C_{CO} C_{O_2}^{0.5} \tag{18}$$

$$r_{H_2\ combustion} = 5.159 \times 10^{15} \exp\left(\frac{-3430}{T}\right) T^{-1.5} C_{O_2} C_{H_2}^{1.5} \tag{19}$$

$$r_{CH_4\ combustion} = 3.552 \times 10^{14} \exp\left(\frac{-15,700}{T}\right) T^{-1} C_{O_2} C_{CH_4} \tag{20}$$

$$r_{water\ gas\ shift} = 2780 \exp\left(\frac{-1510}{T}\right) \left[C_{CO} C_{H_2O} - \frac{C_{CO_2} C_{H_2}}{0.0265 \exp\left(\frac{3968}{T}\right)} \right] \tag{21}$$

The eddy-dissipation reaction rate can be expressed using the following equation:

$$r_{eddy-dissipation} = \alpha_{i,r} M_{w,i} A\rho \frac{\varepsilon}{k} \min\left(\min_R \left(\frac{Y_R}{\alpha_{R,r} M_{w,R}}\right), B \frac{\sum_p Y_p}{\sum_i^N \alpha_{i,r} M_{w,i}} \right) \tag{22}$$

The minimum value of these two contributions can be defined as the net reaction rate.

A1.5.3 Species transport equations

The local mass fraction of each specie Y is computed by using a convection-diffusion equation as follows:

$$\frac{\partial}{\partial t}(\rho Y_i) + \nabla(\rho Y_i \vartheta) = -\nabla \cdot J_i + R_i + S_i \tag{23}$$

where J_i is the diffusion flux of species i due to concentration gradients, R_i is the net generation rate of species i due to homogeneous reaction, and S_i is a source term related to the species i production from the solid heterogeneous reaction. The diffusion flux was computed as a function of the turbulent Schmidt number.

A1.5.4 Heterogeneous reaction rate

Char is the solid devolatilization residue. Heterogeneous reactions of char with the gas species such as O_2 and H_2O are complex processes that involve balancing the rate of mass diffusion of the oxidizing chemical species to the surface of biomass particles with the surface reaction of these species with the char. The overall rate of a char particle is determined by the oxygen diffusion to the particle surface and the rate of surface reaction, which depend on the temperature and composition of the gaseous environment and the size, porosity, and temperature of the particle. The commonly simplified reaction models consider the following overall reactions, char combustion (Eq. 24), H_2O (Eq. 25), and CO_2 char gasification (Eq. 26):

$$C + O_2 \rightarrow CO_2 \tag{24}$$

$$C + H_2O \rightarrow CO + H_2 \tag{25}$$

$$C + CO_2 \rightarrow 2CO \tag{26}$$

The heterogeneous reactions are influenced by many factors, namely, reactant diffusion, breaking up of char, interaction of reactions, and turbulence flow. To include both diffusion and kinetic effects, the kinetic/diffusion surface reaction model (KDSRM) [6, 7] was applied. This model weights the effect of the Arrhenius rate and the diffusion rate of the oxidant at the surface particle. The diffusion rate coefficient can be defined as

$$D_0 = C_1 \frac{\left[\left(T_p + T_\infty \right) \div 2 \right]^{0.75}}{d_p} \tag{27}$$

where d_p means particle diameter.

The Arrhenius rate can be defined as follows:

$$r_{Arrhenius} = A \exp - \left(\frac{E}{R_{T_p}} \right) \tag{28}$$

The final reaction rate weights both contributions and is defined as follows:

$$\frac{d_{m_p}}{d_t} = -A_p \frac{\rho R T_\infty Z_{0X}}{M_{w,0X}} \frac{D_0 r_{Arrhenius}}{D_0 + r_{Arrhenius}} \tag{29}$$

This model was included in the CFD framework by using the UDF tool.

At biomass and air inlets, the mass flow rate was prescribed, while at the outlet the atmospheric pressure level was assigned. At the walls, the materials (steel and insulation) and corresponding thicknesses and heat conduction coefficients were prescribed, as well as the ambient temperature conditions (25°C).

All materials (gas species, solid biomass particles) were assigned appropriate properties from standard thermodynamic tables. The properties of the gas species (density ρ, viscosity μ, thermal conductivity k, and specific heat capacity C_p) were allowed to vary with local main phase temperature, and the mixture value was calculated from its local composition and available Fluent laws (ideal gas law for ρ and mass-weighted mixing law for μ, k, and C_p).

A1.6 Energy conservation

To describe the energy conservation, the following energy conservation equation must be solved for each phase:

$$\frac{\partial}{\partial t}\left(\alpha_q\rho_q h_q\right) + \nabla\cdot\left(\alpha_q\rho_q \vec{v}_q h_q\right) = \alpha_q\frac{\partial}{\partial t}\left(p_q\right) + \overline{\overline{\tau}}_q : \nabla\cdot\vec{v}_q - \nabla\cdot\vec{q}_q + S_q$$
$$+ \sum_{p=1}^{n}\left(\vec{Q}_{pq} + \dot{m}_{pq}h_{pq} - \dot{m}_{qp}h_{qp}\right) \tag{30}$$

The rate of energy transfer is assumed as a function of the temperature difference between phases:

$$\vec{Q}_{pq} = h_{pq}\left(T_p - T_q\right) \tag{31}$$

Provided convection is the main heat transfer mechanism within the reactor, the heat transfer coefficient is related to the Nusselt number of phase q by

$$h_{pq} = \frac{6k_p\alpha_q\alpha_p Nu_q}{d_p^2} \tag{32}$$

Appendix B

B1 Mathematical model

The purpose of this section is to develop a modeling approach able to predict the final composition of the syngas resulting from municipal solid waste (MSW) gasification using numerical simulation. The improved state-of-the art CFD models enable the design and optimization of the gasification processes [8]. The numerical simulation was performed using the CFD solver Fluent based on the finite volume method. The gasification was modeled using Fluent data base for a 2D model and multiphase (gas and solid) model. The solid phase was treated as an Eulerian granular model, while the gas phase is considered as continua. The main interaction between the phases is also modeled, heat exchange by convection, mass (the heterogeneous chemical reactions), and momentum (the drag in gas and solid phase). In the following section, the governing equations will be described.

B1.1 Energy conservation

The energy conservation equation for both phases (gas and solid) is as follows [9]:

$$\frac{\partial}{\partial t}\left(\alpha_q \rho_q h_q\right) + \nabla \cdot \left(\alpha_q \rho_q \vec{v}_q h_q\right) = \alpha_q \frac{\partial}{\partial t}\left(p_q\right) + \bar{\tau}_q : \nabla \cdot \vec{v}_q - \nabla \cdot \vec{q}_q + S_q$$

$$+ \sum_{p=1}^{n} \left(\vec{Q}_{pq} + \dot{m}_{pq} h_{pq} - \dot{m}_{qp} h_{qp}\right) \qquad (1)$$

where $\vec{Q}_{pq}$ is the heat transfer intensity between fluid phase pth and solid phase qth, h_q is the specific enthalpy of phase qth, $\vec{q}_q$ is the heat flux, S_q is a source term due to chemical reactions, and h_{pq} is the enthalpy of the interface.

Eq. (2) [10] describes the rate of energy transfer as a function of the temperature difference between the phases; where the heat transfer coefficient between the phases

$$\vec{Q}_{pq} = h_{pq}\left(T_p - T_q\right) \qquad (2)$$

pth and qth is given by h_{pq}.

The heat transfer coefficient is associated with the Nusselt number of solid phase qth, and k_p is the thermal conductivity for phase pth [8]:

$$h_{pq} = \frac{6 k_p \alpha_q \alpha_p Nu_q}{d_p^2} \qquad (3)$$

Nusselt number is correlated by [10]

$$Nu_s = \frac{h_{gs} d_s}{k_g} = \left(7 - 10\alpha_g + 5\alpha_g^2\right)\left(1 + 0.7 Re_s^{0.2} Pr_g^{0.33}\right)$$

$$+ \left(1.33 - 2.4\alpha_g + 1.2\alpha_g^2\right) Re_s^{0.7} Pr_g^{0.33} \qquad (4)$$

where Re_s is the Reynolds number based on the diameter of the solid phase and the relative velocity and Pr_g is the Prandtl number of the gas phase.

B1.2 Momentum model

The gas and solid-phase momentum equations are as follow: Eq. (5) refers to solid-phase momentum equation, t_s are the particle phase stress tensor and P_s is the particle phase pressure due to particle collisions. Eq. (6) represents the gas-phase momentum equation, where β is the gas-solid interphase drag coefficient, τ_g is the gas-phase stress tensor, and U_s is the mean velocity of solid [9].

$$\frac{\partial}{\partial t}\left(\alpha_s \rho_s \vec{v}_s\right) + \nabla \cdot \left(\alpha_s \rho_s \vec{v}_s \vec{v}_s\right) = -\alpha_s \cdot \nabla p_s + \nabla \cdot \alpha_s \bar{\tau}_s + \alpha_s \rho_s \vec{g} + \beta\left(\vec{v}_g - \vec{v}_s\right) + S_{sg} U_s \qquad (5)$$

$$\frac{\partial}{\partial t}\left(\alpha_g \rho_g \vec{v}_g\right) + \nabla \cdot \left(\alpha_g \rho_g \vec{v}_g \vec{v}_g\right) = -\alpha_g \cdot \nabla p_g + \nabla \cdot \alpha_g \bar{\tau}_g + \alpha_g \rho_g \vec{g} + \beta\left(\vec{v}_g - \vec{v}_s\right) + S_{gs} U_s \qquad (6)$$

B1.3 Mass balance model

The biomass feed changes from solid phase into gas phase by reacting with oxygen, steam, and carbon dioxide. The continuity equations for solid and gas phases are given by Eqs. (7) and (8), respectively [9].

$$\frac{\partial}{\partial t}\left(\alpha_s \rho_s\right) + \nabla \cdot \left(\alpha_s \rho_s \vec{v}_s\right) = S_{sg} \qquad (7)$$

$$\frac{\partial}{\partial t}\left(\alpha_g \rho_g\right) + \nabla \cdot \left(\alpha_g \rho_g \vec{v}_g\right) = S_{gs} \qquad (8)$$

where v is the instantaneous velocity of gas/solid phase, ρ is the density, and α is the volume fraction, the subscripts s denotes the solid phase and subscripts g the gas phase. The mass source term due to heterogeneous reaction, S is expressed by the following equation:

$$S_{sg} = -S_{gs} = M_c \sum \gamma_c R_c \qquad (9)$$

In which R_c is the reaction rate, γ_c is the stoichiometric coefficient, and M_c is the molecular weight. The solid-phase density was assumed to be constant. The gas-phase density was calculated on the basis of ideal gas equation:

$$\frac{1}{\rho_g} = \frac{RT}{p} \sum_{i=1}^{n} \frac{Y_i}{M_i} \qquad (10)$$

where R is the universal gas constant, T is the temperature of the gas mixture, p is the gas pressure, Y_i is the mass fraction, and M_i is the molecular weight of each the species.

B1.4 Turbulence model

A Fluent standard k-ε model was chosen for the turbulence model, as this is the most appropriate model when turbulence transfer between phases plays an important role in gasification in fluidized beds. k is the turbulence kinetic energy and ε is the dissipation rate. They are determined by the next transport equations [1]:

$$\frac{\partial}{\partial t}(\rho k) + \frac{\partial}{\partial x_i}(\rho k_{u_i}) = \frac{\partial}{\partial x_j}\left[\left(\mu + \frac{\mu_t}{\sigma_k}\right)\right] + G_k + G_b - \rho \varepsilon - Y_M + S_k \qquad (11)$$

$$\frac{\partial}{\partial t}(\rho\varepsilon) + \frac{\partial}{\partial x_i}(\rho\varepsilon_{u_i}) = \frac{\partial}{\partial x_j}\left[\left(\mu + \frac{\mu_t}{\sigma_\varepsilon}\right)\frac{\partial\varepsilon}{\partial x_j}\right] + C_{1\varepsilon}\frac{\varepsilon}{k}(G_k + C_{3\varepsilon}G_b) - C_{2\varepsilon}\rho\frac{\varepsilon^2}{k} + S_\varepsilon \quad (12)$$

From Eq. (11), G_k is the generation of turbulence kinetic energy due to the mean velocity gradients, G_b is the generation of turbulence kinetic energy due to buoyancy, and Y_M is the contribution of the fluctuating dilatation in compressible turbulence to the overall dissipation rate. In Eq. (12), $G_k = 1.0$ and $G_\varepsilon = 1.3$ are the turbulent Prandtl numbers for k and ε, respectively, S_k and S_ε are user defined source terms. $G_{1\varepsilon} = 1.44$, $G_{2\varepsilon} = 1.92$, and $G_{3\varepsilon} = 0$ are constants suggested by Launder and Spalding [1].

B1.5 Granular Eulerian model

Granular Eulerian model is described by the following conservation equation for the granular temperature [11]:

$$\frac{3}{2}\left[\left(\frac{\partial(\rho_s\alpha_s\Theta_s)}{\partial t} + \nabla\cdot(\rho_s\alpha_s\vec{v}_s\Theta_s)\right)\right] = (-P_s\bar{I} + \bar{\tau}_s):\nabla(\vec{v}_s) + \nabla(k_{\Theta_s}\nabla\Theta_s) - \gamma_{\Theta_s} + \varphi_{gs}$$

$$(13)$$

This equation is obtained from the kinetic theory of gases. The term $(-P_s\bar{I} + \bar{\tau}_s):\nabla(\vec{v}_s)$ describes the generation of energy by the solid stress tensor, ϕ_{ls} is the energy exchange between the fluid phase and the solid phase, $\gamma_{\Theta a}$ is the collisional dissipation of energy, and $k_{\Theta a}\nabla(\Theta_s)$ is the diffusion energy ($k_{\Theta a}$ is the diffusion coefficient).

The stress in the granular solid phase is achieved by relating the random particle motion and the thermal motion of molecules within a gas accounting for the inelasticity of solid particles. In a gas, the intensity of velocity fluctuation determines the stresses, viscosity, and pressure of granular phase.

B1.6 Chemical reactions model

The chemical reaction rate coefficients are based on the Arrhenius law. Actually, they are empirical and determined by fitting the experimental data. During the devotilization and cracking water shift reaction will occur, the gas species react with the supplied oxidizer and among them. The most common homogenous gas-phase reactions are [12]

$$CO + 0.5O_2 \rightarrow CO_2 + 283\,kJ/mol \quad (14)$$

$$CO + H_2O \rightarrow CO_2 + H_2 + 41.1\,kJ/mol \quad (15)$$

$$CO + 3H_2O \leftrightarrow CH_4 + H_2O + 206.1\,kJ/mol \quad (16)$$

$$H_2 + 0.5O_2 \rightarrow H_2O + 242\,kJ/mol \quad (17)$$

$$CH_4 + 2O_2 \rightarrow CO_2 + 2H_2O + 35.7\,kJ/mol \quad (18)$$

The Arrhenius rates for each one of these reactions can be expressed as follows, respectively [13, 14]:

$$r_{CO\ combustion} = 1.0 \times 10^{15} \exp\left(\frac{-16,000}{T}\right) C_{CO} C_{O_2}^{0.5} \tag{19}$$

$$r_{H_2\ combustion} = 5.159 \times 10^{15} \exp\left(\frac{-3430}{T}\right) T^{-1.5} C_{O_2} C_{H_2}^{1.5} \tag{20}$$

$$r_{CH_4\ combustion} = 3.552 \times 10^{14} \exp\left(\frac{-15,700}{T}\right) T^{-1} C_{O_2} C_{CH_4} \tag{21}$$

$$r_{water\ gas\ shift} = 2780 \exp\left(\frac{-1510}{T}\right) \left[C_{CO} C_{H_2O} - \frac{C_{CO_2} C_{H_2}}{0.0265 \exp\left(\frac{3968}{T}\right)} \right] \tag{22}$$

The eddy-dissipation reaction rate can be expressed using the following equation [15]:

$$r_{eddy-dissipation} = \alpha_{i,r} M_{w,i} A \rho \frac{\varepsilon}{k} \min\left(\min_R\left(\frac{Y_R}{\alpha_{R,r} M_{w,R}}\right), B\frac{\sum_p Y_p}{\sum_i^N \alpha_{i,r} M_{w,i}} \right) \tag{23}$$

The minimum value of these two contributions can be defined as the net reaction rate. The heterogeneous reactions of char (the solid devolatilization residue) with the species O_2 and H_2O are very complex processes. They demand a mass diffusion balance of the oxidizing species at the surface of the biomass particle with the surface reactions of those species with the char. The composition and the temperature of the gases, as well as the temperature, size, and porosity of the particle are important in determining the overall rate of the char.

The most used overall simplified heterogeneous reactions are [12]

$$C + 0.5O_2 \rightarrow CO + 110.5\,kJ/mol \tag{24}$$

$$C + CO_2 \rightarrow 2CO - 172.4\,kJ/mol \tag{25}$$

$$C + H_2O \rightarrow CO + H_2 - 131.3\,kJ/mol \tag{26}$$

The heterogeneous reactions are influenced by many factors, namely, reactant diffusion, breaking up of char, interaction of reactions, and turbulence flow. In order to include both diffusion and kinetic effects, the Kinetic/Diffusion Surface Reaction Model [7] was applied. This model weights the effect of the Arrhenius rate and the diffusion rate of the oxidant at the surface particle. The diffusion rate coefficient can be defined as [15]

$$D_0 = C_1 \frac{\left[(T_p + T_\infty) \div 2\right]^{0.75}}{d_p} \tag{27}$$

The Arrhenius rate can be defined as follows:

$$r_{Arrhenius} = A \exp - \left(\frac{E}{R_{T_p}} \right) \qquad (28)$$

The final reaction rate weights both contributions and is defined as follows [15]:

$$\frac{dm_p}{dt} = -A_p \frac{\rho R T_\infty Z_{0X}}{M_{w,0X}} \frac{D_0 r_{Arrhenius}}{D_0 + r_{Arrhenius}} \qquad (29)$$

This model was included in the CFD framework by using the User-Defined Function tool.

B1.6.1 Pyrolysis

Both homogeneous and heterogeneous reactions are preceded from pyrolysis reactions. Modeling pyrolysis is crucial for MSW gasification purposes.

MSW is thermal decomposed into volatiles, char, and tar. There are several approaches to describe this phenomenon and three main approaches are usually followed: a single-step pyrolysis model, competing parallel pyrolysis, and a pyrolysis model with generation of secondary tar.

In this model, we adopt a pyrolysis model with generation of secondary tar. The MSW is mainly composed by cellulosic and plastic components, where the cellulosic material can be divided in cellulose, hemicellulose, and lignin [16, 17] and the plastics are mainly comprised by polyethylene, polystyrene, and polypropylene, among others.

To distinguish the several components that comprise the MSW, the pyrolysis reactions of cellulosic and plastic groups are considered individually and following an Arrhenius kinetic expression.

The primary pyrolysis equations can be defined as follows:

$$Cellulose \xrightarrow{r1} \alpha_1 volatiles + \alpha_2 TAR + \alpha_3 char \qquad (30)$$

$$Hemicellulose \xrightarrow{r2} \alpha_4 volatiles + \alpha_5 TAR + \alpha_6 char \qquad (31)$$

$$Lignin \xrightarrow{r3} \alpha_7 volatiles + \alpha_8 TAR + \alpha_9 char \qquad (32)$$

$$Plastics \xrightarrow{r4} \alpha_{10} volatiles + \alpha_{11} TAR + \alpha_{12} char \qquad (33)$$

The kinetics for the cellulosic material can be given as follows:

$$r_i = \frac{da_i}{dt} = A_i \exp \left(\frac{-E_i}{T_s} \right) (1 - a_i)^n \qquad (34)$$

where i stands for cellulose, hemicellulose, and lignin (r_{1-3}), A_i is the preexponential factor, E_i is the activation energy, and n is the order reaction. The values for each one of these parameters can be found in Ref. [18]. Average values were considered.

Regarding the kinetic reactions for plastics, data were obtained from Ref. [19] by using the following reactions:

$$r_4 = \left[\sum_{i=1}^{n} A_i \exp\left(\frac{-E_i}{RT} \right) \right] \rho_v \tag{35}$$

where A_i, E_i, and ρ_v are the preexponential factor, the activation energy, and the volatiles density, respectively, and can be found in Ref. [19]. i stands for each one of the plastics that comprise the analyzed MSW.

In this model, it is considered a secondary pyrolysis generating volatiles and secondary tar, as follows:

$$Primary\ Tar \xrightarrow{r5} volatiles + Secondary\ Tar$$

Because, this secondary pyrolysis is also very difficult to treat, a simplified global reaction is used [20]:

$$r_5 = 9.55 \times 10^4 \exp\left(\frac{-1.12 \times 10^4}{T_g} \right) \rho_{TAR1} \tag{36}$$

References

[1] B. Launder, B. Spalding, Lectures in Mathematical Models of Turbulence, Academic Press, London, 1972.

[2] M. Syamlal, T.J. Rogers, MFIX Documentation, Theory Guide, vol. 1, National Technical Information Service, Springfield, VA, 1993.

[3] C. Lun, S.B. Savage, D. Jeffrey, N. Chepurniy, Kinetic theories for granular flow: inelastic particles in couette flow and slightly inelastic particles in a general flow field, J. Fluid Mech. 140 (1989) 223–256.

[4] S. Badzioch, P.G.W. Hawsley, Kinetics of thermal decomposition of pulverized coal particles, Ind. Eng. Chem. Process. Des. Dev. 4 (1970) 521–530.

[5] L. Yu, J. Lu, X. Zhang, S. Zhang, Numerical simulation of the bubbling fluidized bed coal gasification by the kinetic theory of granular flow (KTGF), Fuel 86 (2007) 722–734.

[6] M. Baum, P. Street, Predicting the behavior of coal particles, Combust. Sci. Technol. 3 (1971) 231–243.

[7] M. Field, Rate of combustion of size-graded fractions of char from a low rank coal between 1200K-2000K, Combust. Flame 13 (1969) 237–252.

[8] M. Arnavat, J. Bruno, A. Coronas, Review and analysis of biomass gasification models, Renew. Sust. Energ. Rev. 14 (2010) 2841–2851.

[9] Q. Zhang, L. Dor, W. Yang, W. Blasiak, Eulerian model for municipal solid waste gasification in a fixed-bed plasma gasification melting reactor, Energy Fuel 25 (2011) 4129–4137.

[10] D.J. Gunn, Transfer of heat or mass to particles in fixed and fluidised beds, Int. J. Heat Mass Transf. 21 (1978) 467–476.

[11] S.C. Cowin, A theory for the flow of granular materials, Powder Technol. 9 (1974) 61–69.

[12] A. Demirbas, Hydrogen production from biomass by the gasification process, J. Energy Sources 24 (2002) 59–68.

[13] M. Eaton, D. Smoot, C. Hill, N. Eatough, Components, formulations, solutions, evaluation, and application of comprehensive combustion models, Prog. Energy Combust. Sci. 25 (1999) 387–436.

[14] A. Rouboa, A. Silva, A.J. Freire, A. Borges, J. Ribeiro, P. Silva, et al., Numerical analysis of convective heat transfer in nanofluid, AIP Conf. Proc. 1048 (2008) 819–822.

[15] ANSYS, ANSYS Fluent Theory Guide, Release 150, ANSYS, Inc., 2013.

[16] R.C. Baliban, J.A. Elia, C.A. Floudas, Toward novel biomass, coal, and natural gas processes for satisfying current transportation fuel demands, 1: process alternatives, gasification modeling, process simulation, and economic analysis, Ind. Eng. Chem. Res. 49 (2010) 7343–7370.

[17] O. Onel, A.M. Niziolek, M.M.F. Hasan, C.A. Floudas, Municipal solid waste to liquid transportation fuels—part I: mathematical modeling of a municipal solid waste gasifier, Comput. Chem. Eng. 71 (2014) 636–647.

[18] P. Grammelis, P. Basinas, A. Malliopoulou, G. Sakellaropoulos, Pyrolysis kinetics and combustion characteristics of waste recovered fuels, Fuel 88 (2009) 195–205.

[19] C. Wu, C. Chang, J. Hor, On the thermal treatment of plastic mixtures of MSW: pyrolysis kinetics, Waste Manag. 13 (1993) 221–235.

[20] M.L. Broson, J.B. Howard, J.P. Longwell, W.A. Peter, Products yields and kinetics from the vapor phase cracking of wood pyrolysis tars, AICHE J. 35 (1989) 120–128.

Index

Note: Page numbers followed by *f* indicate figures and *t* indicate tables.